EMIRJES MAZA

STUDY OF HAPLOTYPES OF THE HBB GENE IN SCHOOLCHILDREN IN VENEZUELA

AF671193

EMIRJES MAZA

STUDY OF HAPLOTYPES OF THE HBB GENE IN SCHOOLCHILDREN IN VENEZUELA

HBB GENE HAPLOTYPES

ScienciaScripts

Imprint

Any brand names and product names mentioned in this book are subject to trademark, brand or patent protection and are trademarks or registered trademarks of their respective holders. The use of brand names, product names, common names, trade names, product descriptions etc. even without a particular marking in this work is in no way to be construed to mean that such names may be regarded as unrestricted in respect of trademark and brand protection legislation and could thus be used by anyone.

Cover image: www.ingimage.com

This book is a translation from the original published under ISBN 978-620-0-02147-2.

Publisher:
Sciencia Scripts
is a trademark of
Dodo Books Indian Ocean Ltd. and OmniScriptum S.R.L publishing group

120 High Road, East Finchley, London, N2 9ED, United Kingdom
Str. Armeneasca 28/1, office 1, Chisinau MD-2012, Republic of Moldova, Europe
Managing Directors: Ieva Konstantinova, Victoria Ursu
info@omniscriptum.com

Printed at: see last page
ISBN: 978-620-8-57784-1

Copyright © EMIRJES MAZA
Copyright © 2025 Dodo Books Indian Ocean Ltd. and OmniScriptum S.R.L publishing group

DEDICATION

Firstly, I thank Jehovah my God, for blessing me greatly and giving me the strength and opportunity to achieve one of my greatest dreams.

To my greatest loves, my children Vivian Bernabé, Carlos Alberto and Jeremy Franklyn, I adore them with my soul, they were the impetus to achieve this goal in my life, and for many more things I love them my beautiful children. May this achievement serve as a guide for their studies so that they can improve themselves.

To my great love, my life Franklyn David Figueroa Colón, my beloved husband, thank you for guiding me in my studies, you are a great man, I admire you for being a very intelligent, humble and unconditional person. I love you I love you with my soul.

To my parents, Mirian Cortez De Maza and Jesús Maza for their unconditional love. For having made me the person I am, for guiding me in my studies and supporting me at every moment of my life, I love you with my soul.

To my brothers and sisters, thank you for everything Edison, Edwin, Emerson and Eligia because my joys are also , I love you very much, and my sisters-in-law Glendys, Ana and Carmen, thank you, I love you very much.

I am very grateful to my mother-in-law Hilda Colón for her support, and to my brothers-in-law Carlos, Jorge, Brenda, Glisenis and Mariángeles for helping me when they could, I love them very much.

To all my nephews and nieces, starting with Edwin, Egleidys, Edwmairis, Edigles, Ederson, Cezeus, Edigleinys, Edison Jesús, Irlanda, Naomi, Naolys, Sebastián, Aroncito and Jorgito, I love you very much.

To my friends and comadres (Sara, Patricia, Gladelys, Giliannys, Leomargarit, Dayana, Marilyn), for their different contributions and support, and for always being willing to collaborate at all times. A THANK YOU!

THANK YOU

To the Universidad de Oriente (UDO) and to my tutors, María José González and Martha Bravo, thank you for being there in each of the facets of this research work and for your great collaboration, I thank you for everything you did for me from the bottom of my heart.

To all those working in the Laboratory for Research on Abnormal Haemoglobins (LIHA).

To the directors, teaching staff and workers of the E. B. "Cruz Salmerón Acosta" of Araya and Manicuare for their attention and collaboration. I have no words to thank them for everything they helped me in this stage of my life.

To the parents of the children, for allowing me to take and use the samples for this study.

To Bioanalyst María Figuera for her collaboration in the Clinical Laboratory.

located in the Hospital I "Virgen del Valle" in the municipality of Cruz Salmerón Acosta, Araya.

To my compadre, Licenciado Bioanalista Alexis Urbaneja and Jesús Marchant for their collaboration in taking the blood samples.

To the FONACIT projects MC-2007001066 and MC 2008001053 that funded this research.

To all the teachers who contributed a grain of sand in my time as a student, especially to Prof. Gisela Estrella for each of her words, they were a great impulse for me to mature as a mother, wife and student. Thank you very much, I love you very much.

To all those people who in one way or another accompanied me in this stage of my life. Thank you all very much, may Jehovah bless you.

TABLE OF CONTENTS

SUMMARY

*A study of the frequency of haemoglobinopathies and haplotypes of the HBB gene was carried out in the towns of Araya and Manicuare, Cruz Salmerón Acosta municipality, Sucre state. Between the two populations, a total of 289 peripheral blood samples from school children were analysed. 132 peripheral blood samples were randomly collected in the Araya population and 157 samples in Manicuare. The haematological parameters (Hb, Hcto, CHCM, VCM, HCM) were analysed by an automatic haematological analyser. Peripheral blood samples were then analysed by electrophoresis on cellulose acetate membranes at pH 8.6 to identify the haemoglobin variants present in the schoolchildren, finding 6.2% (n=18) of haemoglobin variants in unrelated individuals and 93.8% (n=289) of individuals homozygous for Hb A. The samples that had variants detected were analysed by HPL-CE for confirmation, reporting the Hb S variant with a frequency of 3.8% (n=5) and the presence of Hb A2↑ with 1.5% (n=2) in the population of Araya, in the same way a frequency of Hb S (3.2%) (n=5) and Hb A2↑ with 3.8% (n=6) in heterozygous form were found in Manicuare. Once the presence of mutations in the HBB gene was confirmed in the two populations, the regions of the HBB gene cluster: 5'ε, $^{G}\gamma$, $^{A}\gamma$, ψβ and 3'ψβ were amplified and each was digested with the specific restriction enzymes HincII and HindIII. Through this analysis, it was possible to identify the characteristics of 7 haplotypes of the HBB*A gene in schoolchildren from the population of Araya, the most frequent being: haplotype 2 with 38.4%, followed by 1 and 10 (15.4%) and an atypical haplotype A (++-+-) with 7.7%. After this, 5 haplotypes of the HBB*A gene were identified in the Manicuare population, with haplotypes 2 being the most frequent (53.8%), followed by haplotypes 1, 11 and an atypical haplotype A (++-+-) with 15.4%. In the case of the HBB*S gene haplotypes identified in the Araya school population, the CAR haplotype was found with 75% and an atypical S haplotype with 25%, and in the Manicuare population the CAR and Ben haplotypes were identified (40%), followed by the Cam haplotype (20%). These results differ from those reported in the Rivero, Bolívar and Benítez municipalities and are in agreement with the results reported in the Sucre municipality. Thus, the Araya and Manicuare populations, Cruz Salmerón Acosta municipality, Sucre state, showed a particular distribution of haplotypes, finding a high frequency for the Bantu haplotype, which means that African individuals from Central Africa were dispersed throughout eastern Venezuela and, therefore, the CAR haplotype is the most frequent in Araya and, in Manicuare, Benin and CAR were found with the same proportion. Suggesting two different histories for the African slave trade and with them the distribution and origin of haemoglobinopathies in eastern Venezuela (Sucre state).*

Keywords: *haplotypes, HBB, sickle cells, sickle cell trait.*

INTRODUCTION

Haemoglobin (Hb) is a heteroprotein in the blood, which is found in red blood cells, transporting oxygen from the respiratory organs to the tissues (Perutz, 1972; Rodríguez and Sáenz, 2005) and carbon dioxide from the tissues to the lungs (Malcorra, 2001; Vives, 2001), and is also involved in the regulation of blood pH in vertebrates (Giardina et al., 2004). Haemoglobin consists of a tetrameric protein structure, which varies structurally according to the stage of embryonic development. In humans, during the first three months of uterine life, three types of embryonic haemoglobins are synthesised, called Gower I, Gower II and Portland (Talmaci et al., 2004; Pérez, 2015). In the fourth month of gestation, the synthesis of foetal haemoglobin Hb F (α_2 γ_2) begins, forming 80% of total haemoglobin in the foetus (Peñuela, 2005; Pérez, 2015).After birth, during the first three months of life, foetal haemoglobin decreases and adult haemoglobin Hb A_1 (α_2 β_2) is synthesised, with a proportion of 97% haemoglobin (Kunkel and Wallenius, 1955; Kazazian, 1990). It is important to note that Hb F will always be present in the organism at 0.5% and Hb A_2 (α_2 δ_2) at 2.5%, making a total of 100% haemoglobin in humans (Malcorra, 2001; Salazar, 2004; Lewis et al., 2008).The genes coding for haemoglobins are located on two different chromosomes forming blocks, the alpha (α) block, located on chromosome 16, which consists of alpha (α) HBA and zeta (ζ) HBZ genes, and the beta (β) HBB gene block, located on chromosome 11, consisting of the epsilon (ε) HBE1, gamma (γ) HBG ($^{G}\gamma$ $^{A}\gamma$), delta (δ) HBD and beta (β) HBB genes (Hoffman et al., 2000; De las Heras and Pérez 2008; Duran et al., 2012; Romero et al., 2015). Haemoglobin is considered the model molecule, because it has contributed substantially to the understanding of the basic principles of genetics and has made it possible to detect and identify mutations in the genes coding for the globin chains of the globin chains of the haemoglobins (Peñuela, 2005; Sáenz, 2005; Brandan et al., 2008; Vera, 2010). These alterations give rise to the most common inherited genetic disorders affecting humans (Pathrapol et al., 2010), such as haemoglobinopathies. They comprise a group of autosomal recessive disorders resulting from point mutations, i.e. the substitution of one DNA nucleotide for , which modifies the genetic code and can induce a change in an amino acid of the resulting globin (structural variants), the most frequent worldwide being haemoglobin S (Hb S), C (Hb C), E (Hb E), H (Hb H) and haemoglobin D punjab or Los Angeles (Hb D- punjab) among others (Martinez et al., 1997; Salazar, 2004; Arends et al., 2007; Bernal et al., 2010).Alterations in haemoglobin can also be the result of mutations that affect the expression of the HBB gene, which cause a total or partial decrease in the synthesis of haemoglobin protein chains known as thalassaemias. The different types of thalassaemia are classified according to the globin chains affected, the most common being α, β and δβ-thalassaemia (Vargas, 2011; Villalba, 2015). Finally, there is hereditary persistence of foetal haemoglobin (PHHF) which is characterised by elevated Hb F synthesis in adulthood (Pérez, 2015). More than 750 haemoglobin variants identified in man have been described, some with clinical significance and others with anthropological significance (Arends et al., 2007; Bernal et al., 2010; Romero et al., 2015).The origin of Hb S is believed to be 30 000 years ago, originating from a point mutation in DNA, resulting in the substitution of adenine for thymine in the HBB gene, leading to a change from glutamic acid to valine at codon 6 of the beta globin polypeptide chain, causing the polymerisation of deoxygenated Hb S in the red blood cell, because glutamic acid is polar and valine is apolar, this decreases the solubility of haemoglobin and gives rise to a gelatinous network of fibrous polymers called tautoids, which agglutinate and deform the red blood cell membrane giving it a sickle or sickle shape, causing its rigidity (Romero et al., 2015). Homozygotes with Hb SS or affected with sickle cell anaemia suffer from a symptomatology consistent with a rapid destruction red blood cells and occlusion of the red blood cell membrane (Romero et al., 2015).of blood vessels, leading to dactylitis, splenic sequestration and bacterial infections in early childhood (first 5 years), painful crises and limb ulcers in adolescence, and kidney and lung damage after the age of 30 (Romero et al., 2015). These symptoms can appear early in life and can fatal if not treated in time (Zafeiriou, 2006; Duran et al., 2012). Hb S carriers are asymptomatic, with basal Hb levels, blood morphology and normal physical development. Hb S concentration is less than 50%; however, in some circumstances of anoxia complications may occur. Hb AS heterozygotes have mild anaemia and under normal circumstances have the same biological efficacy as Hb AA homozygotes with the advantage that sickle cell Hb protects them against the malaria parasite, a phenomenon called compensated polymorphism. In homozygotes (Hb SS), low parasitism by P. falciparum can accentuate haemolysis and increase vasoocclusion and may cause death of the individual early in infancy. Under these conditions, heterozygosity for haemoglobin S confers a selective advantage in regions where malaria is endemic (Pérez, 2015; Villalba, 2015). Sickle cell anaemia and sickle cell trait predominate in the black race, most frequently found in sub-Saharan Africa where the gene is found in up to 90% of the population and in African-American descendants in 10% of the population (Nagel, 1984; Villalba, 2015).Another haemoglobin variant, more frequent worldwide is Hb C, this variant results from a mutation in the same codon 6 of the beta globin polypeptide chain, in this one, the change is from glutamic acid to lysine, hence its cathodic shift at pH 8.6. It is considered to be a mutation independent of Hb S; it appears to have a unique localised origin on the west coast of Africa. Hb C erythrocyte dehydration and intracellular crystal formation (Garcia et al., 2009; Garcia et al., 2010). Homozygous (Hb C) individuals present with symptoms characterised by haemolytic anaemia of mild to moderate intensity with splenomegaly; erythrocyte half-life is decreased and microspherocytes are formed in the circulation. The double heterozygous Hb SC leads to a severe sickle cell disorder; however, it is less severe compared to sickle cell disease. The heterozygote (Hb AC) is asymptomatic and is only discovered in population-based studies (Lewis et al., 2008; Garcia et al., 2010; Villalba, 2015).Recent studies have shown that Hb C also exerts a protective effect against P. falciparum infestation, the finding was made by Fairhurst et al (2005). This was that normal erythrocytes allow parasite binding to adhesion proteins, known as PfEMP- 1 (Plasmodium falciparum Erythrocyte Membrane Protein 1), whereby the red blood cells bind to the walls of blood vessels and cannot be destroyed by the spleen, erythrocytes from heterozygous (Hb AC) or homozygous (Hb CC) individuals specifically compromise this binding. Thus, the haemoglobin C variant also confers a selective advantage in regions where malaria is endemic (Garcia et al., 2010; Villalba, 2015). Within the HBB gene, in addition to the mutations that give rise to the structural variants and thalassaemias, there is also a group of silent mutations known as genetic polymorphisms. The combination pattern of these polymorphisms gives rise to the haplotypes that are inherited together with the Hb S mutation known as the HBB*S gene mutation. Five major haplotypes of HBB*S gene mutation are known, the main ones being: Senegal (Sen), Cameroon (Cam), Benin (Ben), Bantu or CAR and Asian or Arab/Indian (Shimizu et al., 2001), the particular geographical distribution of African haplotypes (Sen, CAR, Cam, Ben) around the world, derived from ancestral migratory patterns across the Sahara desert to the American continent (Pérez, 2015), which could explain why the distribution of the HBB*S gene mutation and malaria is the same, and that in Africa it is due to population movements and natural barriers such as the Sahara desert, which limit the spread of Plasmodium falciparum (Pérez, 2015; Villalba, 2015). The haplotypes of the HBB*S gene have been named according to the geographical area of Africa where they predominate and are used as genetic markers of African origin in population studies. (Arends et al., 2007; Duran et al., 2012;

Villalba, 2015).HBB*S gene haplotypes can be diagnosed by the molecular technique, restriction fragment length polymorphism (RFLP's) Long et al., 1990), depending on the cutting pattern with specific restriction enzymes, the presence (+) or absence (-) of polymorphic sites producing fragments of different lengths can be indicated (De Galiza and Da Silva, 2003). Haplotype characterisation of the HBB gene cluster can be achieved studying five polymorphic sites at 5'ε, $^{G}\gamma$,$^{A}\gamma$, ψβ and 3'ψβ (Sutton et al., 1989; Arends et al., 2007) in DNA amplified by the polymerase chain reaction (PCR) molecular technique, this technique allows the detection of different mutations by using oligonucleotide primers or normal primers and primers containing at their 3' ends mutations complementary to the mutant DNA sequence, by hybridising the DNA to be studied with the mutated primers and the presence of a specific amplification product will allow the identification of the mutation. For this purpose, as many primers as mutations to be studied are used (Bavilaquia et al., 1995; Bravo-Urquiola et al., 2007). importance of haemoglobinopathies in the health field has led to an accelerated development of technologies for their analysis in the laboratory, particularly for rapid and accurate diagnosis (Ruíz, 2003; Peñaloza et al., 2008; Romero et al., 2015). These include methods that allow direct detection of the genomic disorder associated with each of the different haemoglobin molecular alterations (Romero et al., 2015), whose characterisation has made it possible to decipher the cultural and ethnic mosaics of certain populations and to explain, at least partially, the clinical heterogeneity of the different haemoglobin variants (Villalba, 2015). One such method is the biochemical technique, electrophoresis on cellulose acetate membranes at alkaline pH, which allows the separation of haemoglobin variants according to their electrical charge (Tan et al., 1993; Bravo-Urquiola et al., 2004; Calvo-Villas et al., 2006; Colah et al., 2007; García et al., 2009). The cation exchange high performance liquid chromatography (HPLC-EC) technique, which is able to determine the relative percentages of haemoglobins A_1, A_2 and F and any variants of haemoglobin A_1, A_2 and F, is also frequently used. unequivocally, ensuring coverage and improving diagnostic capacity, reducing costs, relative to other techniques (Bravo-Urquiola et al., 2007; Romero et al., 2015). The different alterations in the HBB gene were first described in sub-Saharan Africa and Mediterranean regions by Pauling et al. (1949) and Labie et al. (1984). In 1961, a study by Arends and his group of researchers found a frequency of haemoglobinopathies of 1.6% in Venezuela (Arends, 1984; Arends et al., 2007). For 2007, they reported a frequency of 2.4% of haemoglobin variants in the country (Arends et al., 2007), which represents a slight increase of 0.8%.

According to previous studies, the areas where different abnormal haemoglobinopathies are prevalent in Venezuela are the populations of Tapipa, Miranda state; Campoma, Cariaco, Rivero municipality of Sucre state and the coastal area of Aragua state among others, finding the highest frequency of haemoglobinopathies where the African component is predominant (Arends, 1971; Salazar, 1995; Salazar et al., 1996; Turgeon, 2006; Oropeza et al., 2014).

The presence in Venezuela of the genes that cause the different haemoglobin variants dates back to the 16th century, when at the time of colonisation, the Spanish conquistadors brought Caucasian genes from most European countries (mainly Spain) to America and, through the slave market, black genes from virtually all African countries (Sáenz, 1988).Some slave ships came directly from Africa to Venezuela, and other slaves were brought legal trade transactions with the West Indies, to serve in agriculture and mining work. Runaway slaves also came from Curaçao and Trinidad who remained free in the jungle and were called "black maroons", who formed isolated communities (Arends et al., 2007).

With the arrival of Africans in our country, the S and C variants were introduced. The distribution of these variants seems to follow a pattern closely related to the displacement and establishment of black populations in the country. national. The highest frequencies are found in coastal populations, with an abrupt decrease in populations located in the Andes mountain range (Salazar, 2004; Arends et al., 2007). In 1996, a study was carried out on haemoglobin variants in schoolchildren in Araya and Manicuare, Cruz Salmerón Acosta municipality, Sucre state. In 200 samples collected, a frequency of 2.5% of haemoglobin variants was found in the school population of Araya, and in the school population of Manicuare, a frequency of 0.74% was determined in 135 samples (Betancourt, 1995; Salazar et al., 1996).

In a study conducted on haemoglobinopathy in the population of Guariquen, Benítez municipality, a frequency of 2.0% was determined, and in the school population of Petare, Bolivar municipality, Sucre state, a frequency of 3.0% of structural variants was detected using the HPL-CE technique, with Hb S being the most frequent, followed by Hb C (Medina, 2013). In Sucre state, S and C variants are distributed differently in the different municipalities studied (Arends et al., 1990).

In relation to the studies carried out in Venezuela, on the origin of the haplotypes associated with the HBB gene, we have studied the haplotypes associated with the HBB gene. In the municipality of Rivero in the population of Campoma and Cariaco, 74 chromosomes were evaluated, 70 to study the

HBB*A gene and 4 for the HBB*S gene. In the HBB*A gene, 11 haplotypes were observed, where the most frequent were 2 (37.5%), followed by 3 (20.7%) and 5 (10.2%); the others presented a frequency of less than 10%. The HBB*S gene was found to be associated with the Ben haplotype (50%), followed by the Sen haplotype (25%) and an atypical haplotype (25%) (González et al., 1998). In the municipality of Sucre, Vívenes et al. (2003) found the CAR haplotype (56.8%) with the highest frequency, followed by the Ben haplotype (40.5%) and Sen (2.7%). For western Venezuela, Arends et al. (2007) and Bravo-Urquiola et al. (2007) suggest that in northwestern Venezuela the Ben haplotype is most frequent, followed by CAR and small proportions of the Sen and Cam haplotypes. In the same order of frequency, HBB gene haplotypes were found in the east of the country in some municipalities of the state of Sucre, in 2013 in the research carried out by Medina in the Petare population observed a frequency of the Ben haplotype with 75%, followed by the CAR haplotype (25%) and in Guariquen a frequency of the CAR haplotype was determined with a percentage of 33.3%. These results confirm the introduction of haplobinopathies of African origin in Venezuela during the colonisation process and the subsequent migrations of African individuals (Cunil-Grau, 1987; Salazar, 2004; Bravo-Urquiola et al., 2007).

Due to the racial mixtures in Venezuela, frequency, origin and distribution of the different haemoglobinopathies and haplotypes associated with the HBB gene have been studied throughout the country for several decades (Cunil-Grau, 1987; Vívenes et al., 2003; Salazar, 2004; Bravo-Urquiola et al., 2007; Oropeza et al., 2014).

In this sense, the present research work was carried out with the main objective of determining the haplotypes and associated haplotypes of the HBB gene in schoolchildren and their family group from the Araya and Manicuare populations, Cruz Salmerón Acosta municipality, Sucre state, in the period 2014-2015. The haematological parameters were determined, and the identification of HBB*A and HBB*S gene haplotypes and their association with haemoglobin variants were analysed using molecular techniques, polymerase chain reaction (PCR) and RFLP's; contributing to corroborate the frequency of haemoglobin variants in the Cruz Salmerón Acosta municipality, in accordance with previous studies. With the aim of mapping the distribution of haplotype-associated haplotypes of the HBB gene in the state of Sucre and thus preventing the birth of homozygous children with Hb SS associated with the different African haplotypes.

METHODOLOGY

Study Area

The blood samples were obtained from schoolchildren in the "E. B. Cruz Salmerón Acosta" of Araya and Manicuare (both with the same name), Cruz Salmerón Acosta municipality, Sucre state, Venezuela. The school of Manicuare is located southwest of the Araya Peninsula in the state of Sucre, on the northern shores of the Gulf of Cariaco, bordered on the north by the Gulf of Cariaco, on the south by Cerro Negro, on the east by Tacarigua and on the west by the sector La Vela los Conucos and the school of Araya on the Araya Peninsula, bordered on the north by the Caribbean Sea and the islands of San Pedro de Coche, Cubagua and Margarita; to the south by the Gulf of Cariaco, to the east by the municipality of Ribero, and to the west by the Caribbean Sea (see Figure 1). Blood sampling was carried out in each of schools on different days. Family group samples were collected on the day of genetic counselling, and a gene tree (see Appendix 1) was drawn up for the relatives of the children who were found to be carriers.

Figure 1. Geographical location of the study area.

Parameters haematological parameters

As bioethical guidelines for the sample collection, the sample collectors were provided with The representatives of the children who were evaluated gave informed consent, following the format established by FONACIT code of bioethics and biosafety, for the collection of blood samples from humans (see

Appendix 2). The samples were taken by a health specialist, in this case a bioanalyst.With the collaboration of the specialist, 289 peripheral blood samples were randomly collected by venipuncture from schoolchildren. The samples were collected during one day in each of the schools mentioned above. 132 samples were obtained from the Araya school population and 157 samples were collected from the Manicuare school population using the same procedure. For each blood sample, approximately 5 mL were taken intravenously from the schoolchildren and collected in tubes with 10% sodium EDTA salt as anticoagulant. Once collected, they were transported in an anime cellar with ice to the Clinical Laboratory of the Hospital I "Virgen del Valle" in Araya, Cruz Salmerón Acosta municipality, Sucre state, where the haematological parameters were determined: Haemoglobin (Hb), Haematocrit (Hcto), Mean Corpuscular Haemoglobin Concentration (MCHC), Mean Corpuscular Volume (MCV), Mean Corpuscular Haemoglobin (MCH) by means of an automatic haematological analyser, model Medonic (Sokal and Rohlf, 1969). On the day of the genetic counselling of the family groups of the schoolchildren who were found to be carriers of haemoglobin variants, 15 samples were collected from the family group of the school population in Araya and 18 in Manicuare. A total of 5 families in Araya and 5 in Manicuare had their genetic tree drawn up. Due to the wide variety in age range, school children were separated into children aged 3 to 14 years, while adults in the family group were separated into adult females and adult males aged 15 to 46 years. The normal values for haematological parameters defined by the World Health Organisation (WHO) for children are: Hb 11-16/g- dL, Hcto 35-52%, CHCM 32-36 g/dL, VCM 86-98 FL, HCM 28-39 g/dL, for adult females they are: Hb 12-16/g-dL, Hcto 37-47%, CHCM 32-36 g/dL, VCM 86-98 FL, HCM 28-39 g/dL and for adult males Hb 14-18/g-dL, Hcto 35-52%, CHCM 32-36 g/dL, VCM 86-98 FL, HCM 28-39 g/dL (WHO, 2005).

Electrophoresis on cellulose acetate

Electrophoresis on cellulose acetate membranes at alkaline pH allows the separation of haemoglobin variants according to their electrical charge. In an alkaline medium, most proteins are negatively charged, and when subjected to an electric field, they migrate towards the anode with a speed that is proportional to their net charge. After electrophoresis, the proteins are identified by staining and can be quantified by densitometry (Schneider, 1974). For the electrophoretic run the cellulose acetate membranes were hydrated in a vessel with Tris-EDTA-

Borate buffer (Trismate base 900 mM, EDTA acid 6.3 mM, Boric acid 500 mM) at pH 8.6. The membranes were then removed, excess buffer was carefully removed with filter paper and the membranes were marked for identification. Both compartments of the electrophoresis chamber were filled with 140 mL of H_2O (distilled water) and 20 mL of Tris-EDTA-Borate buffer (TBE) in the (+) cathode compartment and the (-) anode compartment was filled with 150 mL of H_2O and 20 mL of Tris-EDTA-Borate buffer (TBE). After this, strips of filter paper were placed to make a bridge between the buffer and the membrane, facilitating electrical conduction, and the samples made their way through. The blood samples underwent rapid haemolysing, which consisted of adding 90 μL of Saponin and 30 μL of blood to a small, clean glass tube. The applicator was then moistened with the samples in the applicator holder and placed on the acetate membranes previously hydrated with TBE buffer.

The membranes were placed upside down so that the part containing the cellulose was facing down; thus having contact with the bridge strips. The cables of the chamber were connected to the power supply correctly and it was switched on, so that the electrophoretic run was carried out at 250 volts for 15 min. The membranes were then coloured using Ponceau red. To do this, the membranes were placed in a plastic container and the dye was added, left for 10 minutes, then the samples were visualised and haemoglobin variants were detected. After this time, the membranes were washed with 5% glacial acetic acid.The acetate membranes were clarified for preservation, for which they were subjected to a cellulose decomposition process, which consisted of placing the membranes in a container containing methanol and leaving them for one minute, after which the membranes were removed and dried. Subsequently, they were placed in a container containing a mixture of methanol: glacial acetic acid in a 20:80 ratio for one minute, then removed and heated in an oven at 60°C, the membranes became transparent. Once this procedure was completed, the membranes were properly labelled and stored.

High Performance Liquid Chromatography Cation Exchange (HPLC- CE)

Samples for which a haemoglobin variant was detected by cellulose acetate electrophoresis were analysed by HPLC-CE for confirmation using the automated VARIANT Haemoglobin testing system Bio-Rad. This system consists of five programs, one of which, the β-Thalassemia Short program, is able to determine the relative percentages of haemoglobins A_1, A_2 and F (see table 1).HPLC, a technique widely used in biological studies, is based on the separation of molecules according to their retention times. In this case, haemoglobin is adsorbed on an ion exchange column and eluted with solvents of different strength. Given the biochemical differences of haemoglobin variants and their impact on the retention times on the column, these variants can be unequivocally identified.

Peripheral blood samples were prepared by mixing 5 μL of whole blood in 1 mL of haemolysing solution, using a Bio-Rad Autodilutor. Each sample was covered with parafilm and mixed by inversion. They were then placed into the VARIANT along with a series of reagents, which were processed by the equipment at the start of each run. These reagents included a "primer" to condition the column (4.5 x 30 mm) for analysis, a calibrator for haemoglobins A_2 and F and two controls, one normal and one abnormal, designed to monitor the accuracy of the automated quantification (Tan et al., 1993). Each sample was automatically and sequentially added to a pre-programmed gradient of two sodium phosphate buffers of different ionic strength at intervals of 6.5 min. Once in the gradient the samples were transported through the column, where each was separated into its individual components.

The eluted fractions were then passed through a dual wavelength photometric filter (415 and 690 nm), which monitored the elution by detecting changes in absorbance at 415 nm. The 690 nm secondary filter corrected for the effect caused by mixing the buffer with different ionic strength. The absorbance changes were displayed in a chromatogram (absorbance versus time), and the results were printed 3 min after injection.

Table 1. Reference values for normal haemoglobins Tan et al., 1993).

Tipo de hemoglobina	%
A_1	>97,0
F	<0,5
A_2	2,5

Separation of white blood cells by density gradient centrifugation

This technique is used to separate macromolecules according to density, so that heavier molecules precipitate and those of lower molecular weight are suspended, resulting in a separation in bands or zones, to achieve the separation of white blood cells and red blood cells (Mongini and Waldner, 1996). For this, 3 mL of Ficoll-Paque (Ficoll 100 mL, sodium diatrizoate 4005.7 g, calcium edetate disodium 9 g in Pharma Biotech water) was added and 5 mL of the blood sample was carefully added through the walls of the tube with ficoll trying not to mix. After centrifugation, the leucocyte interface was then transferred to another Falcon tube and washed with a TE buffer at pH 7.5 (Tris-HCl 10 mmol. L^{-1}, EDTA 2 mmol. L^{-1}) and resuspended with a shaker (Vortex). Then centrifuged at 3 000 rpm/10 min and the supernatant was discarded. This procedure was repeated four times, washing the cells were resuspended in 1 mL of TE buffer and stored in an eppendorf vial at -20°C.

DNA extraction genomic DNA

The extraction protocol of Welsh and Bunce (1999) was used. The WBC pellet was taken and 9 mL of red cell lysis buffer (RCLB) (NH_4Cl 0.144 mol-L^{-1}, $NaCO_3$ 1 mol-L^{-1}) was added, mixed by inversion several times, allowed to stand and then frozen at -20°C for 30 min, thawed and centrifuged at 3 500 rpm/15 min. The supernatant was discarded, then the pellet was washed with 5 mL of RCLB and centrifuged at 3 500 rpm/15 min. The supernatant was discarded and the pellet was resuspended in 600 µL of white blood cell lysis buffer and sodium dodecyl sulphate (WBC+SDS) (NaCl 0.4 mol. L^{-1}, Tris-HCl 10 mmol. L^{-1} at pH 8.2, disodium EDTA 2 mmol. L^{-1}). The precipitate was then transferred to an eppendorf vial which was left incubating at 56°C overnight to dissolve the precipitate. Once the white blood cell pack was dissolved, 100 µL of NaCl 6 mol. L^{-1} was added and vortexed, then 600 µL of chloroform was added to denature the proteins, vortexed until a milky emulsion was formed and centrifuged again at 11 000 rpm/15 min. The aqueous phase was extracted, and transferred to a new vial; as the aqueous phase was observed to be cloudy, 600 µL of chloroform was added again to denature any proteins that might be present. 1 mL of cold 100% ethanol was added and shaken gently by inversion until a DNA mesh was observed, incubated for 30 min at -20°C and subsequently centrifuged at 11 000 rpm/5 min, the alcohol was discarded by inversion and 1 mL of cold 70% ethanol was added, then centrifuged at 11 000 rpm/5 min. The alcohol was removed and the tube was left to dry by placing it inverted on absorbent paper for 30 min. Once the DNA pellet was dry, 100 µL of ultrapure water was added and incubated for 15 min at 65°C. The DNA was stored for short periods of time at -20°C.

DNA quantification genomic

Quantification of genomic DNA was performed spectrophotometrically at 260 nm, using an eppendorf Biophotometer Plus quantification kit (Kai, 2007). For

A 1/10 dilution of the DNA sample was prepared in a quartz cuvette and the cuvette was introduced into the quantification apparatus to obtain a reading of the concentration of the DNA sample and the degree of purity.

Gel electrophoresis of agarose

Electrophoresis is an analytical technique for the separation of macromolecules, which is based on the differential mobility of charged macromolecules when subjected to an electric field as a consequence of their charge/mass ratio (Westermeier, 1997).This technique was used to estimate DNA quality. For this purpose, a 3% agarose gel was prepared, 1.2 g of agarose was dissolved in 40 mL of 1X TAE buffer (Tris-acetic acid 0.8 mmol. L^{-1}, EDTA 0.02 mmol. L^{-1} at pH 7.5) until a clear solution was obtained. Then 1 μL of ethidium bromide (0.5 μg/mL) was added to visualise the DNA fragments, then allowed to cool and placed in the tray, and the combs were inserted. 3 μL of the DNA sample was taken, mixed by pipetting with 2 μL of loading buffer and deposited into the indentations made with the combs. The electrophoretic run was performed at 150 volts for 15 minutes in 1X TAE buffer. Subsequently, the agar plate was observed in a UV Transilluminator, where the DNA pathway was photographed and evidenced.

Polymerase Chain Reaction (PCR)

PCR allows in vitro enzymatic amplification of specific DNA sequences. The PCR reaction was carried out for a final volume of 50μL, which a 1X Go Taq flexi buffer (50 mmol. L^{-1} Tris-HCL pH 8.3; 100 μg. mL^{-1} gelatin), 1 M magnesium chloride ($MgCl_2$) (3 mmol. L^{-1}), 25 mM dNTP's (0.4 mmol. L^{-1}), 20 pmol of each oligonucleotide (10 mmol. L^{-1}), 2 U of Go Taq flexi Taq polymerase and 100 ng of DNA. This reaction was carried out in a Gene Amp PCR Syatems 2400 thermal cycler (Applied Biosystems) under the following conditions: initial denaturation for 1 min at 94°C, oligonucleotide hybridisation for 1 min at 50-55°C, according to the fragments that were studied and elongation was performed for 2 min at 72°C for 35 cycles. The PCR conditions used for the amplification of polymorphic regions of the HBB gene cluster are represented in (See Appendix 5) according to the protocol of Bavilaquia et al. (1995). Once amplification of the five polymorphic sites was complete, the samples were run on a 3% agarose gel and the gel was stained with ethidium bromide (10 mg.ml-1), run at 150 volts for 15 min and visualised on a UV Transilluminator.

Digestion with restriction enzymes

amplified samples were digested with the HincII and HindIII enzymes to analyse the five polymorphic sites (HincII at 5'ε, HindIII $^{G}\gamma$, HindIII $^{A)}\gamma$, HincII ψβ and HincIII 3'ψβ) (see Figure 2), in order to identify a specific digestion profile in samples. The PCR product was digested with 2 U of each of the HincII and HindIII enzymes, to which 1 µg BSA was added, followed by 2.5 µL of PCR water and the corresponding enzyme. Finally, incubation was carried out at 37°C for 24 h (Long et al., 1990). To visualise the digestion pattern obtained from the fragments amplified with the HincII and HindIII enzymes, 3% agarose gel electrophoresis was performed at 150 volts for 30 min (Westermeier, 1997) and the fragments generated by the polymorphism were visualised using ethidium bromide as a developing agent and a UV Transilluminator. Observation, photographs and storage of the results for subsequent analysis were obtained using a Gel Doc (Bio-Rad) (see Figure 3, 4, 5, 6 and 7).

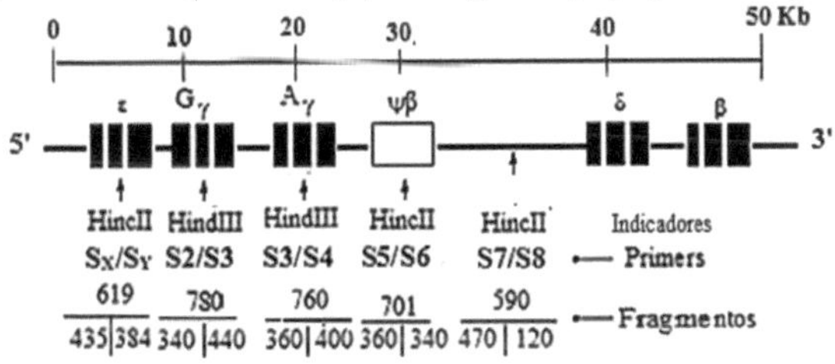

Figure 2. Arrangement of the genes in the HBB block. Boxes represent genes with exons (black) and introns (white). Arrows point to the RFLP`s, which define the haplotypes, as well as the enzymes that recognise them (Long et al., 1990).

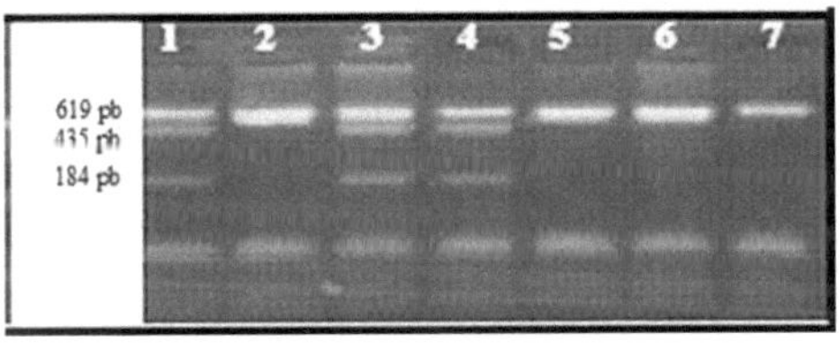

AMP: (+/-) (-/-)

Agarose gel digestion patterns with the HincII enzyme for the HincII 5`ε site. In homozygous schoolchildren with normal Hb AA and heterozygous Hb AS tested in the two populations studied. AMP: amplification.

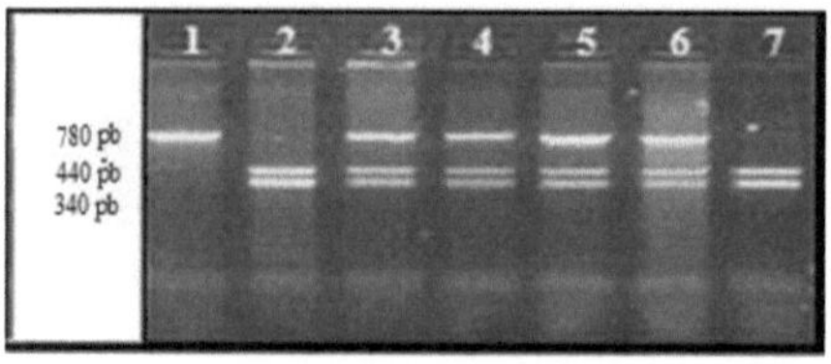

AMP: (-/-) (+/+) (+/-)

Agarose gel digestion patterns with the HindIII enzyme for the $^{G}\gamma$ site. In homozygous schoolchildren with normal Hb AA and heterozygous Hb AS tested in the two populations studied.

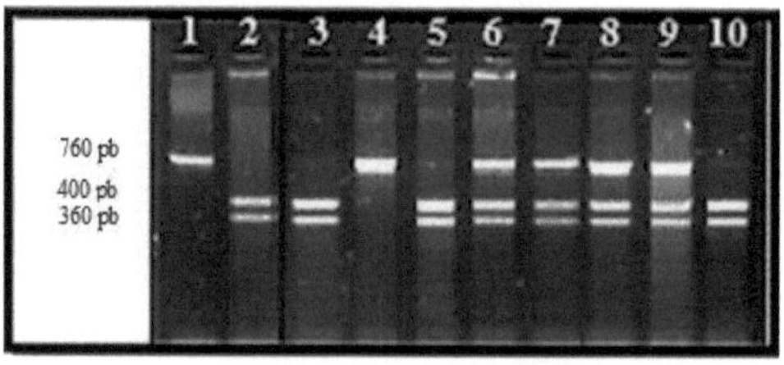

AMP:(-/-) (+/+)(+/-)

Agarose gel digestion patterns with the HindIII enzyme for the Aγ site. In homozygous schoolchildren with normal Hb AA and heterozygous Hb AS tested in the two populations studied.

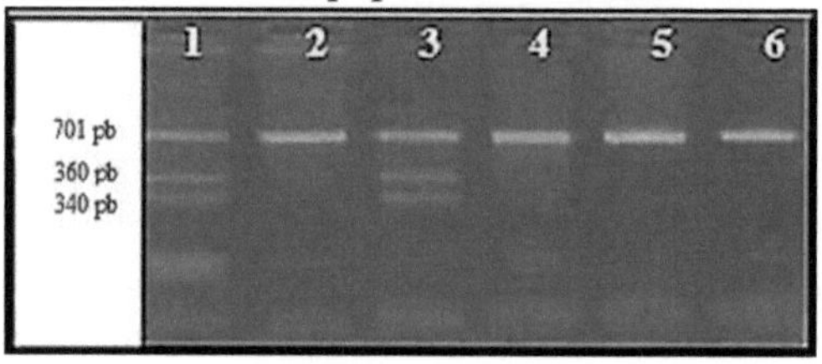

AMP:(+/-) (-/-)

Agarose gel digestion patterns with the HincII enzyme for the HincII $\psi\beta$ site. In schoolchildren heterozygous with Hb AS and homozygous $A_2\uparrow$ and homozygous with Hb AA in the two populations studied.

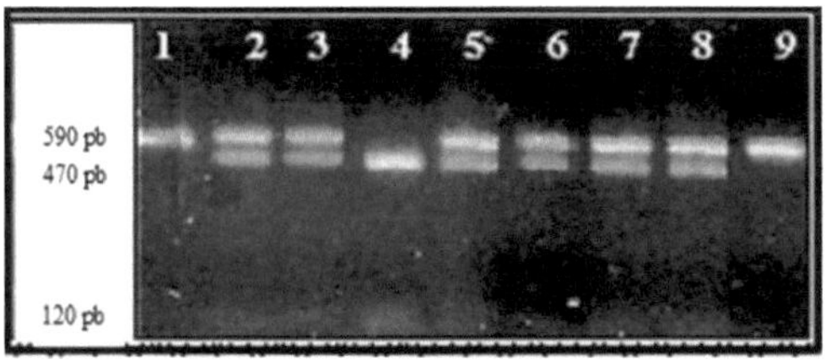

AMP: (-/-) (+/-)(+/+)

Agarose gel digestion patterns with the HincII enzyme for the HincII 3'ψβ site. In schoolchildren heterozygous with Hb AS, heterozygous $A_{(2)}$↑ and homozygous with Hb AA in the two populations studied.

Haplotype assignment

For haplotype assignment, the restriction enzyme cut sites on the amplified DNA were analysed. The presence or absence restriction sites were denoted by the signs (+) or minus (-) respectively according to the classification of (Long et al., 1990). Haplotypes that could not be classified by this nomenclature were designated as atypical (Sutton et al., 1989).

Analysis of data

The determination of the frequency and association between haemoglobin variants and haplotypes of the HBB*A and HBB*S gene with haematological parameters were expressed in tables and figures.

RESULTS

We analysed 289 peripheral blood samples obtained from schoolchildren from the "E. B. Cruz Salmerón Acosta" of Araya and Manicuare, using the cellulose acetate membrane electrophoresis technique at alkaline pH as shown in Appendix 3, in order to identify the haemoglobin variants present in the populations studied. We found a frequency of 6.2% (n=18) of haemoglobin variants and Hb A in homozygous state with 93.8% (n=271) between the two populations (see Figure 8).

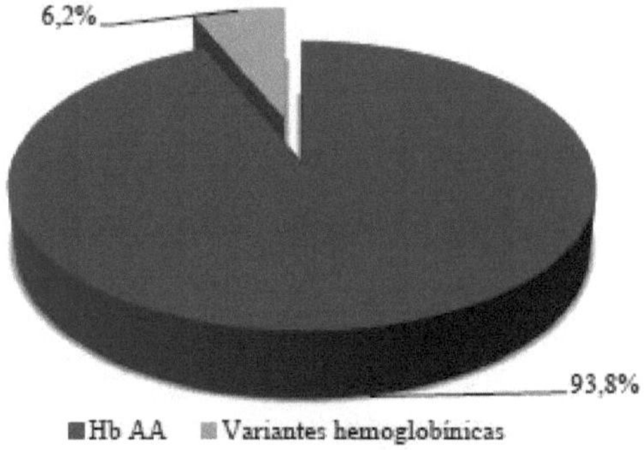

Frequency of haemoglobin variants in unrelated schoolchildren from Araya and Manicuare, Cruz Salmerón Acosta municipality, Sucre state, detected cellulose acetate membrane electrophoresis at alkaline pH. The genotypes present were Hb AA, Hb AS and Hb $AA_{(2)}\uparrow$. For confirmation, they were analysed by HPL-CE, since Hb S and D when exposed to an electric field migrate towards the anode with the same speed proportional to their net charge; by analysing the HPL-CE technique, it was possible to determine that the S variant was present and thus concordance was found between the two techniques used.Therefore, in the Araya population, Hb AA was identified in homozygous form with the highest frequency of 94.7% (n=125), it was found associated with the S variant with a frequency of 3.8% (n=5) and with Hb $A_2\uparrow$ with 1.5% (n=2) in heterozygous form (Hb AS and Hb A_2 with 1.5% (n=2) in heterozygous form). Hb $AA_{(2)}\uparrow$) (see Figure 9).Figure 10 shows that in the school population of Manicuare, out of 157 samples analysed, 146 individuals presented Hb AA in homozygous form corresponding to a frequency of 93%, it was also found associated with the S variant with 3.2% (n=5) and with Hb $A_2\uparrow$ with 3.8% (n=6) heterozygous form (Hb AS and Hb $AA_{(2)}\uparrow$).

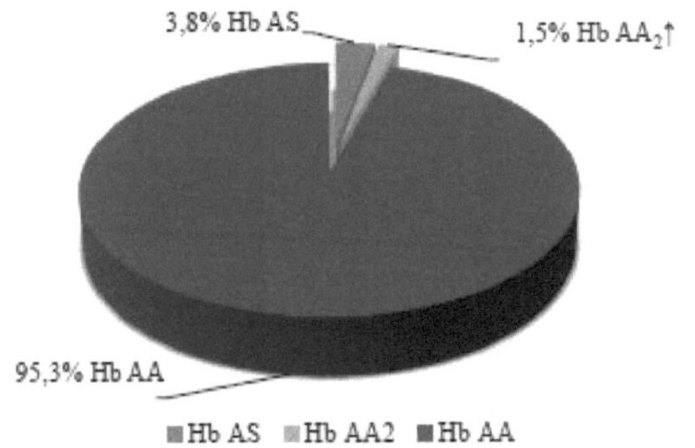

Frequency distribution of haemoglobin variants detected by cellulose acetate membrane electrophoresis in the population of Araya, Cruz Salmerón Acosta municipality, Sucre state.

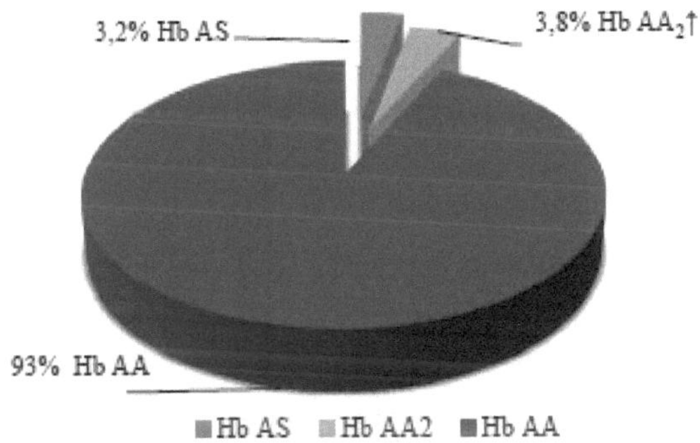

Frequency distribution of haemoglobin variants detected by cellulose acetate membrane electrophoresis in the population of Manicuare, Cruz Salmerón Acosta municipality, Sucre state.

In individuals with sickle cell trait, Hb A_1 values quantified by HPLC-CE ranged between 58-59%, Hb A_2 between 2.5-5.5% with a slight increase, Hb F between 0.2-0.4% and the structural variant found (Hb S) ranged between 37.8-38.1% (see Appendix 4). In all cases studied, no false positive cases were found. In this sense, it was observed that the Araya and Manicuare populations had the same frequency of sickle cell trait.

The haematological parameters and their values are shown in Table 2, with the phenotypes of boys, schoolgirls and the family group with the phenotypes of adult females and adult males of the Araya population, determined by age and sex.

Table 2. Haematological parameters of schoolchildren and the family group of the population of Araya, Cruz Salmerón Acosta municipality, Sucre state.

	Araya		Parámetros hematológicos				
Ind.	Núm de casos	Tipos de Hb (%) (×±DS)	Hb (g/dL) ($\bar{x}$±DS)	HCTO (%) ($\bar{x}$±DS)	VCM (FL) ($\bar{x}$±DS)	HCM (pg) ($\bar{x}$±DS)	CHCM (g/dL) ($\bar{x}$±DS)
		Fenotipo AA					
Niños	125	93±1	12±1	34±1	80±3	32±1	32±2
Adultos hembras	3	97±2	9,5±1	35±3	87±9	33±2	37±3
Adultos varones	1	90	15	38	93	39	37
		Fenotipo AS					
Niños	7	53±2	13±3	28±2	79±2	28±2	32±6
Adultos hembras	1	55	9	27	77	23	31
Adultos varones	4	56±2	14±2	30±2	83±1	35±3	26±2
		Fenotipo AA_2↑					
Niños	2	83±2	12±1	21±2	69±2	18±1	26±2
Total	143						

Haemoglobin (Hb), Mean Corpuscular Volume (MCV), Mean Corpuscular Haemoglobin Concentration (MCHC), Mean Corpuscular Haemoglobin (MCH), Arithmetic Mean (X), Standard Deviation (SD), Individual (Ind).

The schoolchildren in the Araya population with sickle cell trait had Hb values of 12.2±3 g/dL and HCTO 34±1 %. One of the AS carriers represented by an adult female presented decreased values of haematological parameters (Hb 9 g/dL, VCM 77 FL, CHCM 31 g/dL, HCM 23 pg, HCTO 27 %). The 2 schoolchildren with Hb $AA_{(2)}$↑ had decreased values of MCV 69±2 FL, MCHC, 26±2 g/dL, MCH 18±1 pg, HCTO 21±2 %. With respect to the haematological parameters of the school children of Manicuare, the AA homozygotes presented Hb 13.2±1 g/dL, MCV 75±3 FL, MCHC 36±2 g/dL, MCH 27±1 pg, HCTO 32±1%. While schoolchildren with sickle cell trait presented haematological parameters Hb 12.1±3 g/dL, HCTO 35±3 %, MCV 77±6 FL, MCH 25±5 pg, HCTO 33±1 g/dL. Importantly, individuals with Hb $A_{(2)}$↑ were found to have decreased haematological parameters (Hb 11±1 g/dL, HCTO 19±2 %, MCV 69±2 FL, MCHC 21±2 g/dL, MCH 15±1 pg), Similarly, two adult females with sickle cell trait were found with decreased erythrocyte values (Hb 9,5±3 g/dL, HCTO 35±2 %, HCM 33±2 pg, VCM 57±2 FL, CHCM86±2 g/dL) see Table 3). In the analysis of HBB gene polymorphisms, a total of 18 individuals were studied. Twenty-six chromosomes were analysed for the HBB*A gene and 10 for the HBB*S gene in unrelated individuals. We were able to identify 7 haplotypes, 6 typical and 1 atypical in 13 chromosomes for the HBB*A gene in the Araya population, the most frequent being 2 (+).38.4%, followed by 1 (- - - - - - - -) and 10 (+ + + - - + + + +) 15.4%, haplotypes 5 (- + + - - + + + +), 12 (+ + + - - - -), 14 (+ + + - - - +) and an atypical haplotype A with a frequency of 7.7% (see table 4). The HBB*S gene was found to be associated with the CAR haplotype (-+) with a higher frequency (see Table 4). frequency of 80% in heterozygous form and with an atypical haplotype S (++ -+ -) with 20% (See Table 5).In the Manicuare population, 5 haplotypes were determined, 4 typical and 1 atypical in 13 chromosomes for the HBB*A gene, in which haplotypes 2 (+ - - - - - -) were determined with a higher frequency (53.8%), followed by 1 ( ), and an atypical haplotype (+ - - - - -) with a higher frequency (+ - - - - -), followed by 1 ( ), and an atypical haplotype (+ - - - - -) with a higher frequency (+ - - - - - -). + --) with 15.4% and 5, 8, 9 (- ++++) and 11 (- - - - -+ +) with a lower percentage 7.7% (See Table 7). The HBB*S gene was found to be associated with the Bantu or CAR haplotype (-+ - -) with 40% and Ben (+) with the highest frequency with a percentage of 40%, followed byof the Cam haplotype (20%) (see Table 8).

Table 3. Haematological parameters of the schoolchildren and the family group of the population of Manicuare, Sucre state.

			Parámetros hematológicos				
Ind.	Núm de casos	Tipos de Hb (%) (×±DS)	Hb (g/dL) ($\bar{x}$±DS)	HCTO (%) ($\bar{x}$±DS)	VCM (FL) ($\bar{x}$±DS)	HCM (pg) ($\bar{x}$±DS)	CHCM (g/dL) ($\bar{x}$±DS)
		Fenotipo AA					
Niños	146	89±1	13,2±1	32±1	75±3	27±1	36±2
Adultos hembras	5	86±1	12,3±2	34±3	92±9	29±1	35±3
Adultos varones	3	91±1	14±1	33±1	91±7	30±3	32±2
		Fenotipo AS					
Niños	5	51±2	12,1±3	35±3	77±6	25±5	33±1
Adultos hembras	3	57±2	11,3±3	36±1	86±2	33±2	36±2
Adultos varones	3	53±2	13,5±2	32±2	86±1	38±3	26±4
		Fenotipo AA_2↑					
Niños	6	84±2	11±1	19±2	69±2	15±1	21±1
Total	171						

Haemoglobin (IIb), Mean Corpuscular Volume (MCV), Mean Corpuscular Haemoglobin Concentration (MCHC), Mean Corpuscular Haemoglobin (MCH), Arithmetic Mean (X), Standard Deviation (SD), Individual (Ind).

In the Araya population, a haplotype was found most frequently with a combination pattern (2/CAR) in 2 individuals, in one individual with the haplotypes, the haplotypes (10/CAR) in 1 individual, one individual with the haplotypes (Atypical A/CAR) and finally one individual with the haplotypes (1/Atypical S). In Manicuare, the CAR haplotype was found most frequently in heterozygous AS individuals, in 1 case 2/CAR, in 1 individual 1/CAR and in 1 case 8/CAR. While the Benin haplotype presented a combination pattern in 1 individual 2/Ben and in 1 case 8/CAR. individual 2/Cam. The Asian haplotype was not found in the samples analysed AND the Senegal haplotype. Frequency of haplotypes of the HBB*A gene studied in schoolchildren from Araya, Cruz Salmerón Acosta municipality, Sucre state.

Restriction Sites

Haplotypes	HincII	HindIII	HindIII HincII		HincII	Chromosome No.	Freq (%)
	5'ε	G γ	A γ	ψβ	3'ψβ		
1	-	-	-	-	-	2	15,4
2	+	-	-	-	-	5	38,4
5	-	+	-	+	+	1	7,7
10	+	+	-	+	+	2	15,4
12	+	+	-	-	-	1	7,7
14	+	+	-	-	+	1	7,7
Atypical	+	-	-	+	-	1	7,7
Total						13	100

*Designation according to Long et al. (1990).
*Designation according to Villalobos et al. (1997).

Distribution of haplotypes of the HBB*S gene studied in AS heterozygous schoolchildren from Araya, Cruz Salmerón Acosta municipality, Sucre state.

Restriction Sites

Haplotypes	HincII 5'ε	HindIII G γ	HindIII HincII A γ	ψβ	HincII 3'ψβ	Chromosome No.	Freq (%)
Bantu or CAR	-	+	-	-	-	4	80
Atypical	-	+	-	+	-	1	20
Total						5	100

*Designation according to Sutton et al. (1989).

The haematological parameters and HBB gene haplotypes determined in individuals with sickle cell trait in both populations studied are shown in Table

8. Individuals with sickle cell trait with CAR or Bantú de Araya haplotypes presented normal haematological values.

Frequency of haplotypes of the HBB*A gene studied in schoolchildren from Manicuare, Cruz Salmerón Acosta municipality, Sucre state.

Restriction Sites

Haplotypes	HincII 5'ε	HindIII HindIII Gγ	Aγ	HincII ψβ	HincII 3'ψβ	Chromosome No.	Freq (%)
1	-	-	-	-	-	2	15,4
2	+	-	-	-	-	7	53,8
8	+	-	-	+	+	1	7,7
11	-	-	-	+	+	1	7,7
Atypical	+	-	-	+	-	2	15,4
Total						13	100

*Designation according to Long et al. (1990).
*Designation according to Villalobos et al. (1997).

Table 7. Distribution of haplotypes of the HBB*S gene studied in AS heterozygous schoolchildren from Manicuare, Cruz Salmerón Acosta municipality, Sucre state.

Restriction Sites

Haplotypes* HincII HindIII HindIII HincII HincII HincII **No. of Chromosome Frec** (%)

5'ε	Gγ	Aγ	ψβ	3'ψβ		
Bantu or CAR -	+	-	-	-	2	40
Benin (Ben) -	-	-	-	+	2	40
Cameroon (Cam) -	+	+	-	+	1	20
Total					5	100

*Designation according to Sutton et al. (1989).

Of the 15 samples collected in the family group of the Araya school population, 8 individuals were AS heterozygotes with a frequency of 53.3% and 43.7% (n=7) were AA homozygotes. In the case of the Manicuare family group, out of 18 samples collected, 6 were positive for Hb S with a frequency of 33.3% and 12 individuals were homozygous for Hb AA with a frequency of 66.7%.

A genetic polymorphism study was carried out on the family group of schoolchildren who were found to be carriers of Hb S. For this diagnosis, 16 chromosomes were analysed for the HBB*A gene and 14 for the HBB*S gene. Four typical haplotypes could be identified in 10 chromosomes for the HBB*A gene in the Araya family group, the most frequent being 2 (+ - - - - - -) 70%, followed by 1 ( ), 8 (+ - - -+ +) and 10 (++ -+ +) 10% (See Table 9). The HBB*S gene was found to be associated with the CAR haplotype (- + - - - - -) with the highest prevalence with a frequency of 100% in heterozygous form (See Table 10). With a combination pattern in 6 individuals 2/CAR, one individual 8/CAR and one individual 10/CAR. Five haplotypes, four typical and one atypical in 13 chromosomes for the HBB*A gene were determined in the family group of the schoolchildren of the Manicuare population, in which haplotypes 2 (+) were determined with the highest frequency (53.8%), followed by haplotype 2 (+), followed by haplotype 2 (+) and haplotype 3 (+).1 (- - - - - - - - -), and an atypical haplotype (+- - -+ -) with 15.4% and 5, 8, 9(- ++++) and 11 (- - - - - - - -).+ +) with a lower percentage of 7.7% (see Table 11). The HBB*S gene was found to be associated with the Bantu or CAR haplotype (-+) with 50% of the highest frequency, followed by the haplotypeBen (- - - - - - +) with 33.3% and Cam (16.7%) (see Table 12). With a combination pattern in 2 individuals 2/CAR, one individual 9/CAR, one individual 2/Ben, one individual 5/Ben and one individual 11/Cam. Figures 11, 12, 13, 14 and 15 show the family trees of families carrying the S allele and the haplotypes of the HBB gene studied in the Araya population.

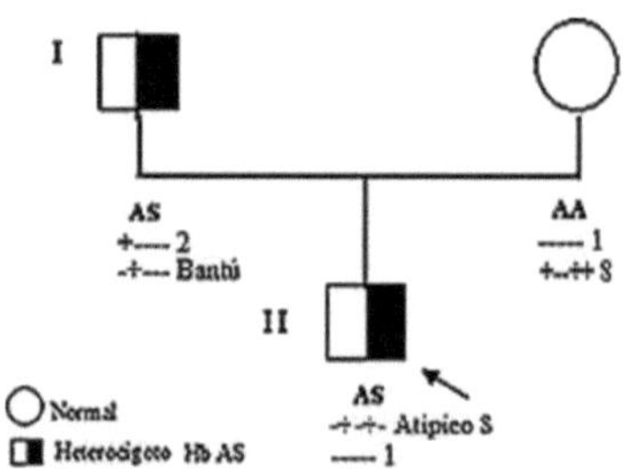

Figure 11. Genealogical tree of family I carrying the S allele and the associated haplotypes of the HBB gene studied in the population of Araya, Sucre state.

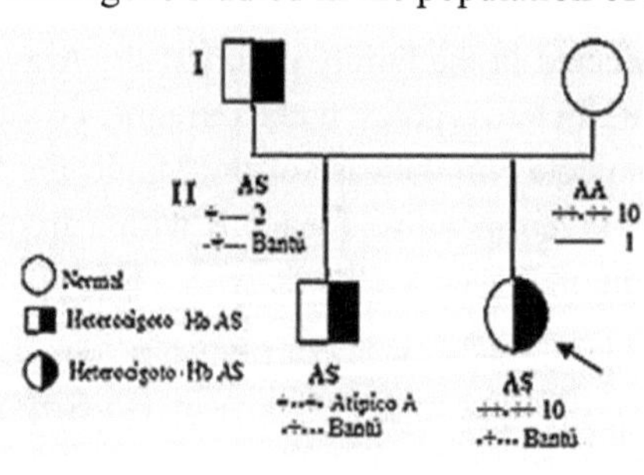

Figure 12. Genealogical tree of family II carriers of the S allele and associated haplotypes of the HBB gene studied in the population of Araya, Sucre state.

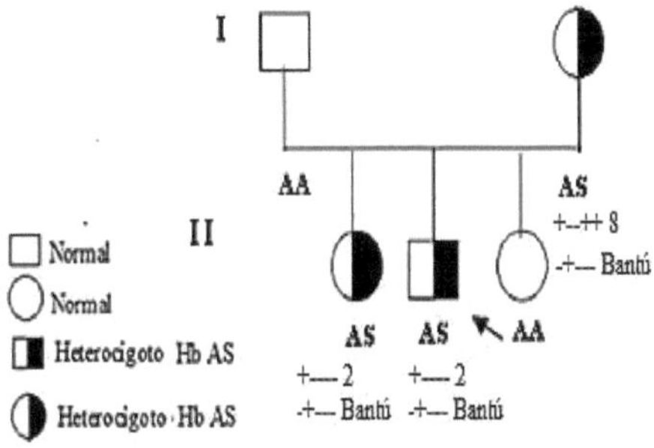

Figure 13. Genealogical tree of family III carriers of the S allele and associated haplotypes of the HBB gene studied in the population of Araya, Sucre state.

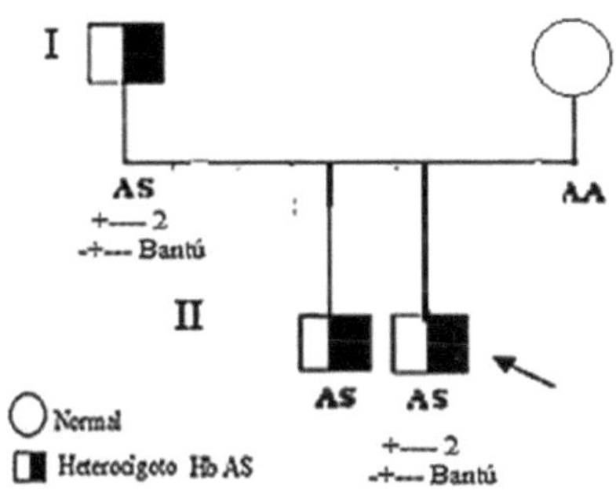

Figure 14. Genealogical tree of family IV carriers of the S allele and associated haplotypes of the HBB gene studied in the population of Araya, Sucre state.

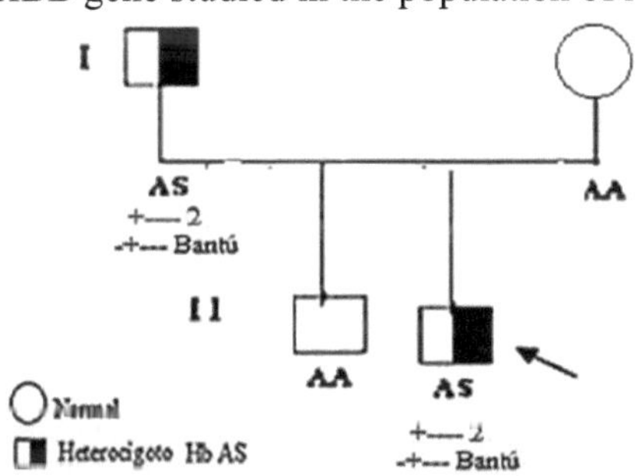

Figure 15. Genealogical tree of the family V carriers of the S allele and the associated haplotypes of the HBB gene studied in the population of Araya, Sucre state.

Haematological parameters and haplotypes of the HBB*S gene determined in AS heterozygous schoolchildren from Araya and Manicuare, Cruz Salmerón Acosta municipality, Sucre state.

Araya Haematological parameters

No. of cases	Haplotypes	Phenotype	Hb (g/dL) (× ± SD)	HCTO (g/dL) (× ± SD)	VCM (FL) (× ± DS)	MCHC (g/dL) (× ± SD)	HCM (pg) (× ± SD)
3	CAR or Bantu	AS	12±1	34±1	89±3	33±2	32±1
1	Atypical S	AS	14,1	34,3	77	27,8	35,9
	Manicuare						
2	CAR or Bantu	AS	13±2	33±2	79±3	37±2	35±1 2
2	Benin (Ben)	AS	12±1	22±3	79±1	30,7±1	38,9±1
1	Cameroon (Cam)	AS	13,6±1	33,3±1	77±1	14,8±1	36±1

Haemoglobin (Hb), Mean Corpuscular Volume (MCV), Mean Corpuscular Haemoglobin Concentration (MCHC), Mean Corpuscular Haemoglobin (MCH), Arithmetic Mean (X), Standard Deviation (SD).

Figures 16, 17, 18, 19 and 20 show the family trees of families carrying the S allele and haplotypes of the HBB gene studied in the Manicuare population.

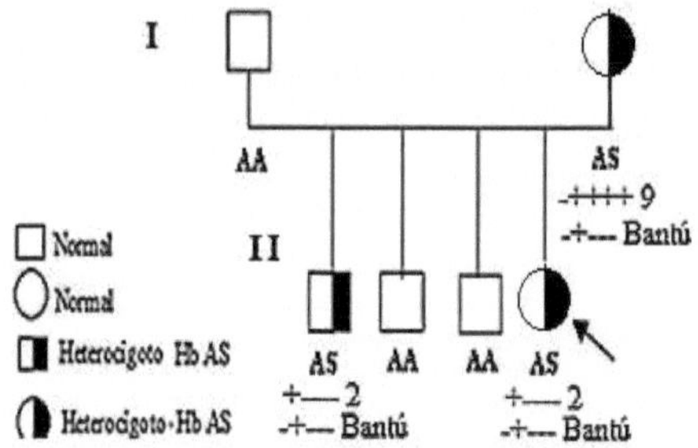

Genealogical tree of family I carrying the S allele and the associated haplotypes of the HBB gene studied in the population of Manicuare, Sucre state.

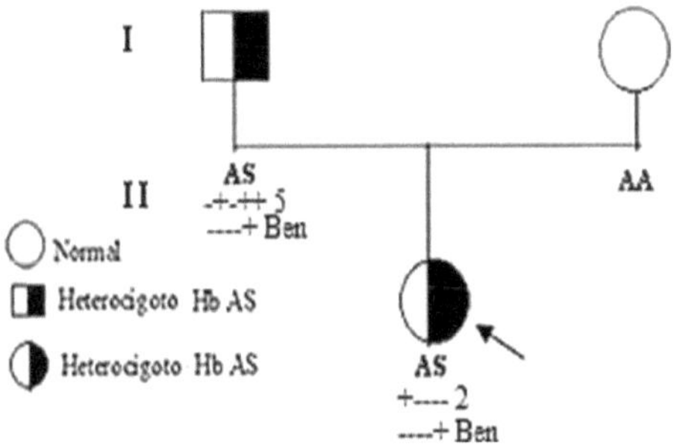

Figure 17. Genealogical tree of family II carrying the S allele and the associated haplotypes of the HBB gene studied in the population of Manicuare, Sucre state.

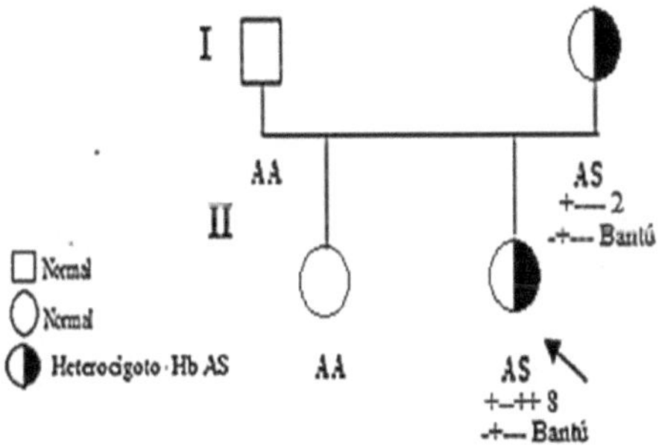

Figure 18. Genealogical tree of family III carriers of the S allele and haplotypes of the HBB gene studied in the population of Manicuare, Sucre state.

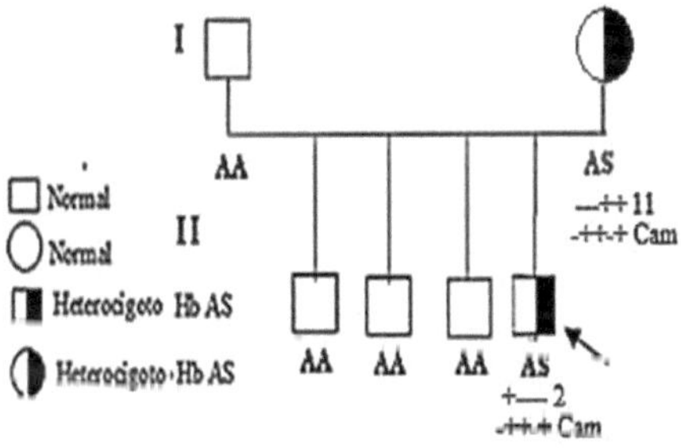

Figure 19. Family tree of family IV carriers of the S allele and associated haplotypes of the HBB gene studied in the population of Manicuare, Sucre state.

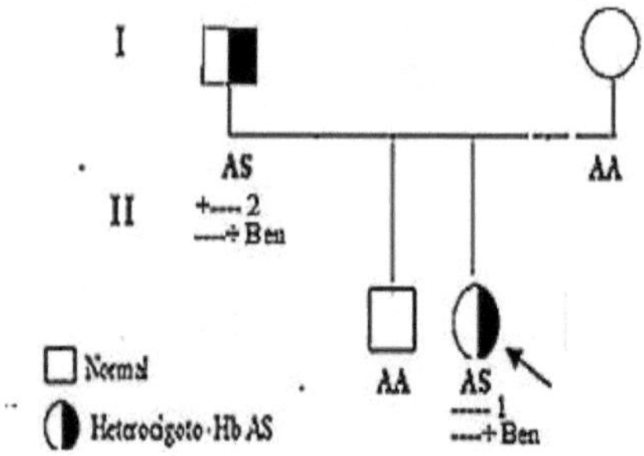

Figure 20. Genealogical tree of the family V carriers of the S allele and the associated haplotypes of the HBB gene studied in the population of Manicuare, Sucre state.

Table 9. Frequency of haplotypes of the HBB*A gene studied in the family group of schoolchildren from Araya, Cruz Salmerón Acosta municipality, Sucre state.

Restriction Sites

Haplotypes	HincII	HindIII	HindIII HincII		HincII	Chromosome No.	Freq (%)
	5'ε	G γ	A γ	ψβ	3'ψβ		
1	-	-	-	-	-	1	10
2	+	-	-	-	-	7	70
8	+	-	-	+	+	1	10
10	+	+	-	+	+	1	10
Total						10	100

*Designation according to Long et al. (1990).

Table 10. Distribution of haplotypes of the HBB*S gene studied in the family group of schoolchildren from Araya, Cruz Salmerón Acosta municipality, Sucre state.

Restriction Sites

Haplotypes* HincII HindIII HindIII HindIII HincII HincII HincII **No. of Chromosome Freq** (%)

	5'ε	Gγ	A γ	ψβ	3'ψβ		
Bantu or CAR	-	+	-	-	-	8	100
Total						8	100

*Designation according to Sutton et al. (1989).

Frequency of haplotypes of the HBB*A gene studied in the family group schoolchildren from Manicuare, Cruz Salmerón Acosta municipality, Sucre state.

Restriction Sites

Haplotypes	HincII 5'ε	HindII I G γ	HindII I A γ	HincI I ψβ	HincII 3'ψβ	No. of Chromosome	Freq (%)
2	+	-	-	-	-	3	50
5	-	+	-	+	+	1	16,7
9	-	+	+	+	+	1	16,7
11	-	-	-	+	+	1	16,7
Total						6	100

*Designation according to Long et al. (1990).

Table 12. Distribution of haplotypes of the HBB*S gene studied in the family group of schoolchildren from Manicuare, Cruz Salmerón Acosta municipality, Sucre state.

Restriction Sites

Haplotypes* HincII HindIII HindIII HindIII HincII HincII HincII **No. of Chromosome Freq** (%)

5'ε^{G} γ^{A} γψβ3'ψβ

Bantu or CAR -	+	-	-	-	3	50
Benin (Ben) -	-	-	-	+	2	33,3
Cameroon (Cam) -	+	+	-	+	1	16,7
Total					6	100

*Designation according to Sutton et al. (1989).

DISCUSSION

From the analysis of 289 peripheral blood samples, it was possible to determine the presence of haemoglobin variants of African origin in both populations. According to the comparison of the frequency of haemoglobin variants found among the schoolchildren belonging to the Araya and Manicuare schools, it was possible to determine the highest frequency for the S variant with 3.8% (n=5) and 1.5% (n=2) for the S variant,8% (n=5) and 1.5% (n=2) for Hb $A_{(2)}\uparrow$ in Araya, in contrast to the variants found in Manicuare, where 5 individuals with variant S had 3.2% and 3.8% of Hb $A_{(2)}\uparrow$ (n=6). In this sense, it was observed that the Araya and Manicuare populations had the same frequency of sickle cell trait, which could generate a serious public health problem in these two populations in the future.The presence of Hb S of African origin in Araya and Manicuare is partly due to the geographical location or the dynamism between the human transit of African and European individuals during the colonial period, to which both populations were subjected in past centuries. According to Navarrete, 2005, the Araya peninsula was the point of arrival of slave ships and one of the important centres of the slave trade, because the ships that brought African slaves for forced labour landed on the Araya peninsula, such as the Dutch who brought African slaves to the peninsula to work in the salt mines of the Araya peninsula.These results are consistent with the results obtained by Arends et al. (2007) in a study on haemoglobinopathies carried out in Venezuela; in this study, 80 400 samples of individuals from different regions of the were analysed, finding 9% (7 305 individuals) of haemoglobinopathies, of which 80% were structural variants in heterozygous, homozygous or associated with a minor thalassaemia. Similarly, Medina (2013) reported a frequency of 2.0% of haemoglobin variants in the Benítez and Bolívar municipalities, Sucre state, in Guariquen, Benítez municipality, and 3.0% of variants in Petares, Bolívar municipality, with the S variant being the most , followed by C variant. The results obtained in Araya and Manicuare in schoolchildren and their family group in the period 2014-2015, corroborate the results obtained in a study carried out in Araya and Manicuare on haemoglobin variants by Salazar et al. (1996), in this year they found the presence of haemoglobin variants with a frequency of 2.5% of the analysis in 200 peripheral blood samples, finding one sample with Hb C and another with Hb $SA_{(2)}\uparrow$, two presented values of Hb $A_{(2)}\uparrow$, one of them with evidence of β-thalassaemia minor and one sample presented increased foetal Hb in the school population of Araya, and in Manicuare only one was heterozygous AS with a frequency of 0.74% (n=135), these results led these researchers to conclude that the variants in these two populations are sporadic or the result of resent migrations (small migrations), since the individuals who had the variants detected in Manicuare came from the Rivero municipality.As can be seen, the high frequency of Hb S of African origin and Hb A_2↑ as indicative of beta thalassaemia minor generally in Caucasian individuals from the Mediterranean reported in Araya and Manicuare in the 2014-2015 school period. It is due to the high rate of inbreeding to which these two populations have been subjected for a long time, as all individuals with sickle cell trait found in these two populations reported being inhabitants born in these two populations, and in the family study the elders mentioned being founders of Araya and Manicuare, which means that these haemoglobin variants are not sporadic in these populations and are due to the large migrations at the time of the colony.Once again, the presence of haemoglobin variants in the populations of the state of Sucre and their intimate relationship with the arrival of Europeans and Africans during the colonisation process is evident; the distribution of these variants responds to a pattern determined by the way in which these populations settled in Venezuelan territory (Acosta, 1967; Cunil-Grau, 1987; Sáenz, 1988; Arends et al., 2000; Salazar et al., 2002; Arends et al., 2007). It is known that the state of Sucre has a 16% African component in its population (Salazar, 2004).Consequently, the manifestation of Hb A_2 ↑ found in two individuals from the Araya population with decreased values of MCV 69±2 FL and MCH 18±1 pg and 6 in Manicuare with MCV 62±2 FL and MCH 15±1 pg respectively, this could mean that these values lead these individuals to a microcytosis with hypochromia (MCV and MCH) below the normal limit. Thus, elevated Hb A_2 levels are variable according to the mutation that causes this disorder and are generally considered indicative of β-thalassaemia minor or "thalassaemic trait" in Caucasian individuals from the Mediterranean (Evans et al., 2003; Villalba, 2015). This means that in the Manicuare population there is a great variability of races, with the European component being found most frequently, followed by the African component.Salazar (2004) points out that, by federal entities, the highest incidence of Hb S in Venezuela is found in the states of Bolívar, Guárico, Cojedes and Aragua, the latter being the state with the highest percentage of this genetic alteration. In turn, the states of Táchira, Mérida and Nueva Esparta have the lowest frequency of this haemoglobin variant. In the rest of the states, the frequency of haemoglobin S ranges between 1.5 and 3.5%. As for the haematological parameters determined in schoolchildren from both populations studied, the average values of Hb, Hct, MCV, MCCH and MCH in schoolchildren with the AA phenotype did not indicate haematological alterations and the values remained within the reference range proposed by the WHO (WHO, 2005).In the cases of Hb S carriers diagnosed in the two study populations, they showed minimal alterations in the haematological series, this may be due to the fact that more than half of their synthesised blood is genetically altered, while the other half is normal, which would help to counteract the effects that the mutation could cause in these individuals, exhibiting the same biological efficacy as normal homozygotes (Hb AA) (Sebastiani et al., 2005; Pérez, 2015; Villalba, 2015), which is consistent with the results recorded by Salazar et al. (1996) in schoolchildren from the Araya and Manicuare populations.The exception was with three AS carriers represented by three adult females who had decreased values of haematological parameters (Hb, MCV). They also had decreased basal Hb during pregnancy, which led to iron deficiency anaemia. The three individuals with sickle cell trait showed hypoxia when they were in confined spaces, for example in the sea transport known as "Tapaitos", which makes it difficult for these people to travel constantly in this type of transport Therefore, haematological parameters may vary to some extent depending on the environment in which the individual develops, on genetic factors or in many cases on elements involved in other diseases that compromise the health of the individual (Medina, 2013).With regard to the analysis of HBB*A gene polymorphisms, 12 haplotypes were identified. The haplotype with the highest frequency in both populations studied was haplotype 2 (+). The same haplotype was reported in oceanic populations, such as Melanesia and Polynesia. They were also reported in Brazilian tribes and in other Venezuelan populations. For Vivenes et al. (2003) it was the most relevant haplotype found between the coasts of Sucre and Anzoátegui. Haplotype 2 (+) of the HBB*A gene is the second most frequent gene in Mexican populations with African roots. Its presence in Araya and Manicuare is an indication of admixture with Oceanic, European, Amerindian and African populations. According to Long et al. (1990) haplotype 2 is the result of an ancestral or other first order haplotype, originated by mutations or gene conversions.In Araya and Manicuare, the same atypical haplotype (+ + + - +-) reported by Villalobos et al. (1997) was found in a Mexican mestizo indigenous population. This same haplotype was reported by González et al. (1998), in the Afro-American population of Campoma, Sucre state.The finding of atypical haplotypes of the HBB*A and HBB*S gene in Araya and Manicuare is the product of several recombination events between ancestral haplotypes that were fixed as a result of linkage disequilibrium, which is indicative of relatedness between closely related individuals. This may cause the haplotypes present on the ancestral chromosomes for sickle cell disease (CAR or Bantu, Ben, Cam, Sen and Arab/Hindu) and the haplotypes of the HBB*A gene chromosome, recombine in heterozygous individuals (for the haplotypes) causing their progeny to display atypical haplotypes (Villalobos et al., 1997; Medina, 2013). In the HBB*S gene polymorphism study, the Bantu haplotype (- + - - - -) was obtained with the highest frequency in the Araya population (80%), the same haplotype that predominates in individuals from the states of Sucre, Anzoátegui, Cojedes and Carabobo (Vivenes et al., 2003; Arends et al., 2007). It was also found in Brazil, Colombia, Guadeloupe and throughout the Caribbean. These regions were involved in arrival of slave ships from Africa to the colonies in the Americas during the height of colonisation (Duran et al., 2012; Romero et al., 2015; Villalba, 2015).In the Manicuare population, an order of frequency of the HBB*S gene haplotypes was determined as , the Bantu haplotype (-+ - - - -) and Ben (+) with a 40%, followed by the Cam haplotype (- + + + - - +) with a percentage of 20%. It is important to note that the Sen and Arab/Hindu haplotypes were absent in both populations. Thus, the same order of frequency of haplotypes found in several studies in Brazil and Colombia was found. Moreno et al. 2002 in Aragua State and Arends et al. (2007) reported that in northwestern Venezuela the Ben haplotype is the most frequent, followed by CAR and small proportions of Sen and Cam. The family study was carried out in order to corroborate the origin of the S allele found in the school populations studied. Thus, Hb S was determined more frequently in both populations. In the Araya family group, 53.3% of the S variant was found in 8 individuals. In the case of the Manicuare family group 6 were positive for Hb S with a frequency of 33.3%. In the Manicuare family group, the CAR haplotype was found most frequently, followed by Ben and Cam. However, in Araya, only the CAR haplotype was reported in the family group.The family trees were carried out in order to identify the individuals who The gene trees of the Araya family groups show that most of them inherited the allele from the father and one group inherited it from the mother. As could be seen in the gene trees of the family groups in Araya, the majority inherited the allele from the father and one group inherited it from the mother. In Manicuare 3 children inherited the S-allele from the mother and 2 children inherited it from the father. In this way it was easier to explain to each family group the possibilities of having children homozygous for Hb S and the haplotypes associated with this mutation.With the study of genetic polymorphism in Araya and Manicuare, it was possible to demonstrate the population of Africans in the Cruz Salmerón Acosta municipality, Sucre state, and that the majority of these individuals had their origin in the Central African

Region (CAR), and small proportions in Benin and Cameroon. These results differ from those reported for the Rivero, Bolívar and Benítez municipalities, which presented an order of frequency for the HBB*S gene haplotypes (Ben, CAR and Sen) (González et al., 1998; Medina, 2013). Suggesting two different histories for the slave trade in eastern Venezuela (Sucre state) and with them a different distribution of haemoglobinopathies and their haplotypes associated with the HBB gene. In several studies, it has been postulated that the highest proportion of individuals who came to America from Africa originated in Benin and small proportions in the Central African Region (CAR), Senegal and Cameroon (González et al., 1998; Moreno et al., 2002; Arends et al., 2007; Medina, 2013). However, in our finding, the CAR haplotype was found with a higher proportion in both populations studied.An important difference between the present study and others related to haplotypes and haplotypes of the HBB gene in Venezuela lies in the populations studied. Each population showed a particular distribution of haplotypes, which explains why the distribution of haplotypes in the populations included in this study suggests that individuals from central Africa (Bantu) were dispersed throughout eastern Venezuela and, therefore, contributes the most frequent haplotype in Araya and Manicuare, Cruz Salmerón Acosta municipality, Sucre state. The identification at the molecular genetic level of polymorphisms of the HBB*S gene have The cultural and migratory patterns of African, African-American and Asian peoples have been studied. In the Americas, it has been possible to decipher the exact origins of black slaves over the past centuries. The evident interbreeding of the population makes haplotypes and haemoglobin variants a useful tool to corroborate historical information on the settlement of Venezuela (Arends, 1971; Salazar, 2004; Bravo-Urquiola et al., 2007). Thus, the high frequency of mixed haplotypes found in the country could influence the clinical behaviour of cases (Bravo-Urquiola et al., 2007).Individuals who presented haplotypes CAR (-+ -), Benin (+), Cam (- ++ +)and Atypical (++ -+ -) S, showed minimal alterations in haematological parameters, because Hb S carriers are only affected on one allele of the gene. However, they generally do not show changes in haematological parameters.Apparently there are no iron deficiency anaemias, nor abnormalities in the rest of the haematological parameters in the populations studied; the apparent absence of deficiency anaemia is related to the dietary habits of these people where their main sustenance is fish as a source of animal protein, which in addition to being rich in iron (heme) favours the absorption of the non-heme present in the vegetable products of the diet (Cook and Monsen, 1976). Although it should be noted that there are studies that show that iron intake and haemoglobin levels in the blood are independent variables (Mata et al., 1993). Some studies have reported that the most accentuated haematological alterations are observed in patients with sickle cell anaemia or sickle cell trait who inherit CAR (- + - - - - -), Benin (+) and Cam (-+ + - +) haplotypes, in which they present amore severe symptomatology, in contrast to those who inherit Sen haplotype (-+ -+ +) and Arab/Hindu who have intermediate severity, this is due to the persistence of Hb F after foetal age in the individual in some cases (Duran et al., 2012; Villalba, 2015). Although this relationship is not always present, the haplotypes of an individual allow an approximation to be made of the severity of sickle cell disease. In addition, haplotypes represent an indicator of the response that may express an individual with sickle cell disease or sickle cell trait to treatment with hydroxyurea, the most common treatment for the disease.

Hydroxyurea increases Hb F concentrations, but in patients with Bantu and Cam haplotypes, hydroxyurea concentrations close to the toxic range of the drug are necessary to achieve an Hb F concentration that significantly minimises disease symptoms. Additionally, the response to the drug decreases rapidly over time in these patients (Duran et al., 2012). Early knowledge of an individual's haplotypes allows action to be taken regarding the type of treatment that should be used to minimise harm. This is an appropriate occasion to insist on the real and true public health problem that sickle cell diseases represent in several countries or regions of our America. Except in the exceptional cases of hypoxia in Hb AS heterozygous individuals, since it is a benign condition that does not constitute a notable public health problem, although it is a problem of education and genetic counselling. Sickle cell disease is considered a public health problem by the WHO due to its high prevalence in certain regions worldwide, especially in underdeveloped countries where management is still inadequate, with a lack of screening, prevention and control programmes, and late diagnosis when the patient already has irreversible complications.The primary objective should focus on improving the care of this group of patients, seeking an early and comprehensive approach, implementing mass screening programmes in school populations, so as to prevent the arrival of children with sickle cell anaemia when schoolchildren decide to have their ancestry in the future. For this reason, this research was carried out to determine the haemoglobin variants and associated haplotypes of the HBB gene in schoolchildren and their families from Araya and Manicuare, Cruz Salmerón Acosta municipality, Sucre state. Hb S carriers were made aware of how delicate and serious anaemia is. The parents and relatives were given several lectures to inform them about sickle cell disease. They were then given genetic counselling to prevent the birth of new homozygous children. These studies are important because they represent a contribution to the knowledge of the genetic structure of the Venezuelan population, in this case the population of the state of Sucre, in the Cruz Salmerón Acosta municipality. This study is part of a group of investigations being carried out in the state of Sucre with the aim of identifying the municipalities with the highest incidence of HBB gene haplotypes associated with different haemoglobinopathies. The relationship between the different haplotypes of the HBB*S gene and the heterogeneous symptoms of sickle cell anaemia has made it possible to elucidate, albeit only partially, the biological expression of this disease, which must be properly diagnosed with appropriate techniques to rule out other factors capable of masking the true haemoglobin genotype of the individual.

CONCLUSIONS

In this research, the presence of Hb A in homozygous form and the S and A_2 variants↑ in heterozygous form was detected in both populations with a high frequency.The origin of the Hb S detected in the Araya and Manicuare populations once again highlights the existence of the different racial stocks reported above. The high levels of Hb A_2 reported in some of the Araya and Manicuare schoolchildren could infer that mutations of Mediterranean origin such as thalassaemias occurred in these localities, maintaining an important European component in these populations.AS heterozygous schoolchildren did not show alterations in haematological parameters, probably due to the geographical conditions of the areas studied. School children and AS heterozygous representatives had decreased Hb, MCV and MCH probably due to the presence of the Bantu and Benin haplotypes which confer complications under adverse conditions to individuals with sickle cell trait.Haplotype 2 of the HBB*A gene was found more frequently in the two study populations, probably due to the result of an ancestral or other first-order haplotype, originating from mutations or gene conversions in both populations. The presence of two atypical haplotypes for the HBB*A and HBB*S gene could be due to the various recombination events between ancestral haplotypes.The presence of the CAR or Bantu haplotype of the HBB*S gene with a higher frequency in Araya is an that this population is closely related to regions of the Central African Republic. Suggesting two different histories for the slave trade in colonial times and with them the distribution and origin of haemoglobinopathies in eastern Venezuela (Sucre state). The Manicuare population presented the same frequency ratio with Benin and Bantu regions, which means that this is a population with a great variety of African ancestry, taking into account the presence of the Cameroon haplotype, which until now has been the only population in Sucre state that has reported this haplotype.The studies on haplotypes of the HBB gene represent a contribution to the knowledge of the genetic structure of the Venezuelan population.The presence of haemoglobinopathies in Araya and Manicuare could be due to the large migrations during the colonisation process.

RECOMMENDATIONS

The communication channel between the diagnostic laboratory, hospital centres and relatives of schoolchildren found to be Hb S carriers should be optimised so that more and more family groups are studied.It is recommended that the Venezuelan government invest more budget in research projects related to population-based studies to detect haemoglobinopathies.Consideration should be given to including other health centres, so that individuals have better access to testing for different haemoglobinopathies. Because sickle cell anaemia is a very common disease in the general population. The focus should be on improving care for this research group, seeking an early and comprehensive approach, implementing mass screening programmes in school populations, in order to prevent the arrival of children with sickle cell anaemia.

APPENDICES

GENETIC TREE FAMILIA

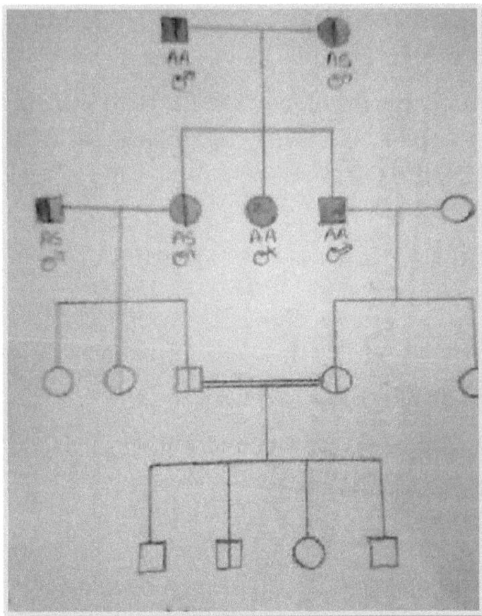

BROTHERS 123

1. Genetic tree carried out on relatives of children who were found to be carriers.

INFORMED CONSENT

PROJECT TITLE: Study of haplotypes of the HBB gene in schoolchildren in the Cruz Salmerón Acosta municipality, Sucre state, 2014-2015.
PROJECT COORDINATOR: Prof. María José González and Dr. Martha Bravo Urquiola.

TELEPHONE: 0293.40.25.29 - 0414.383.85.99

INTITUTIONS:

- Universidad de Oriente, Núcleo de Sucre, Department of Biology.
- Central University of Venezuela.
- LIHA (Laboratory for the Investigation of Abnormal Haemoglobins).
- José Izquierdo" Anatomical Institute.

OBJECTIVES OF THE STUDY:

- To determine the haematological parameters: Haemoglobin (Hb), Haematocrit (Hcto), Mean Corpuscular Haemoglobin Concentration (MCHC), Mean Corpuscular Volume (MCV), Mean Corpuscular Haemoglobin (MCH) in school children and their family group from the villages of Araya and Manicuare, Cruz Salmerón Acosta municipality, Sucre state.
- To identify haemoglobin variants in schoolchildren and their families from the population of Araya and Manicuare, Cruz Salmerón Acosta municipality, Sucre state, by electrophoresis on cellulose acetate membranes.
- Quantify the relative parameters Hb A_1, A_2, F and haemoglobin variants by cation exchange high performance liquid chromatography (HPLC-EC).
- To identify the haplotypes of the HBB gene using the molecular technique fragment length polymorphism of the HBB gene.

restriction (RFLP's).

- Assign HBB gene haplotypes according to restriction enzyme cut sites on PCR-amplified DNA.

The information and biological materials obtained in this study will be used solely and exclusively for purposes of this research.I:C.I

Being of full age and of sound mind and without any compulsion on my part, I declare:

- Have been informed in a clear and simple manner of all aspects related to this study.
- That the team conducting the study has assured me of confidentiality concerning my identity and all information relating to me.
- That under no circumstances have I been offered, nor do I intend to receive, any financial benefit from the findings of the study.
- That the main benefit I will get will be to receive the diagnostic report.

DECLARATION OF THE VOLUNTEER:

After having been informed and having clarified my doubts regarding the study, I agree to participate in the present study.

Signature of volunteer:

Date:

Signature of the researcher:

Date:

Name of enumerator:

2.Informed consent for the collection of blood samples

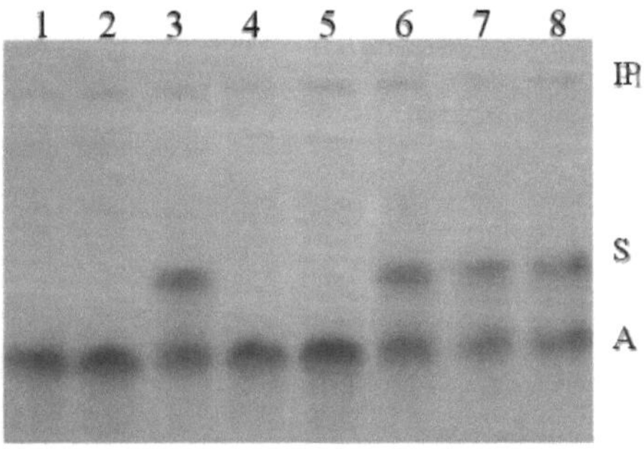

PI: Starting point of the run

3. Electrophoresis of haemoglobin in cellulose acetate, at pH 8.6, to show the migration pattern of the different haemoglobins. In lane 8 is the control (AS); 1, 2 and 4 normal (AA); 3, 6 and 7 heterozygous AS.

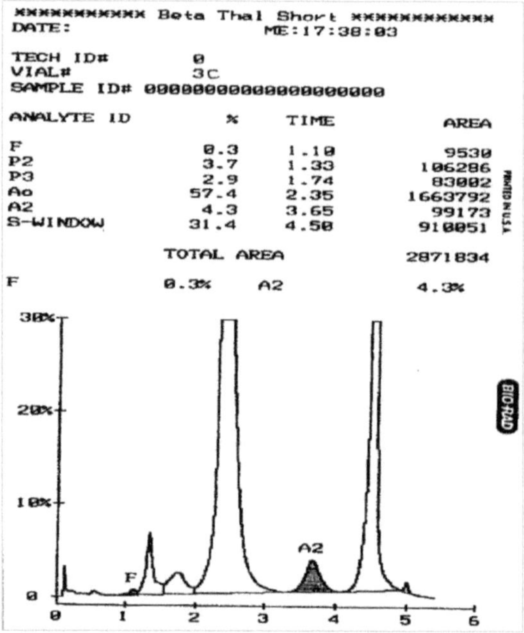

4.Chromatogram showing the presence of a heterozygote with haemoglobin AS.

Indicadores "Primers"	Oligonucleótidos	Inicial T (°C)	Inicial Tiempo	Desnaturalización T (°C)	Desnaturalización Tiempo	Alineamiento T (°C)	Alineamiento Tiempo	Elongación T (°C)	Elongación Tiempo	Extensión final T (°C)	Extensión final Tiempo
Sx Sy	TAGTCCCACTGTGGACTACTT CCTGAGAGCTTGCTAGTGATT	94	1'	94	15"	55	15"	72	23"	72	90"
S2 S3	AAGTGTGGAGTGTGCACATGA TGCTGCTAATGCTTCATTACAA	94	1'	94	15"	60	15"	72	23"	72	90"
S3 S4	TGCTGCTAATGCTTCATTACAA TAAATGAGGAGCATGCACACAC	94	1'	94	15"	60	15"	72	23"	72	90"
S5 S6	GAACAGAAGTTGAGATAGAGA ACTCAGTGGTCTTGTGGGCT	94	1'	94	15"	55	15"	72	23"	72	90"
S7 S8	TCTGCATTTGACTCTGTTAGC GGACCCTAACTGATATAACTA	94	1'	94	15"	55	15"	72	23"	72	90"

(Header spans: "Desnaturalización" above Inicial; "35 Ciclos" above Elongación / Extensión final.)

5. PCR conditions used for the amplification of polymorphic regions of HBB gene cluster according to the protocol of Bavilaquia et al. (1995).

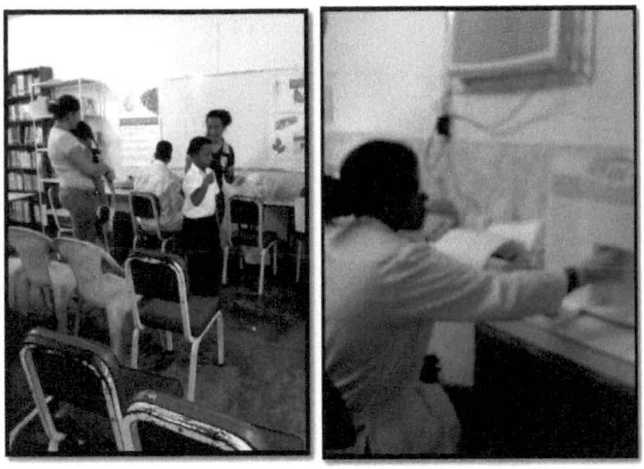

6.Blood sampling day for schoolchildren at the E.B. "Cruz Salmerón Acosta", Araya, Cruz Salmerón Acosta municipality, Sucre state.

7.Blood sampling day for schoolchildren at the E.B. "Cruz Salmerón Acosta", Manicuare, Cruz Salmerón Acosta municipality, Sucre state.

8. Sampling of the family group of schoolchildren who were found to be carriers of Hb AS, Araya, Cruz Salmerón Acosta municipality, Sucre state.

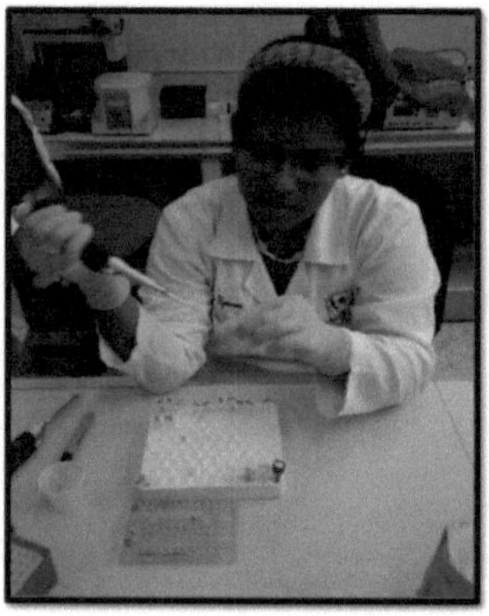

9. Analysis of polymorphism in the HBB gene in schoolchildren and the family group of Araya and Manicuare, Cruz Salmerón Acosta municipality, Sucre state.

BIBLIOGRAPHIES

Acosta, M. 1967. Life of the black slaves in Venezuela. Hesperides. Caracas Venezuela.

Arends, T. 1961. The problem of haemoglobinopathies in Venezuela. Revista Venezolana de Sanidad y Asistencia Social, 26: 61-68.

Arends, T. 1971. Epidemiology of Hemoglobin Variants in Venezuela. In: Genetical, Functional and Physical Studies of Hemoglobin. Arends, T.; Bemsky, G.; Nagel, R. (Eds.) Karger. Brasilea, Switzerland. Pp. 509-559.

Arends, T. 1984. Epidemiology of haemoglobin variants in Venezuela. Gaceta Médica Caracas, 92: 189-224.

Arends, T.; Salazar, R.; Anchustegui, M. and Garlin, G. 1990. Hemoglobin Variant in the Northeastern Region of Venezuela. Interciencia, 15: 36-41.

Arends, A.; Álvarez, M.; Velázquez, D.; Bravo, M.; Salazar, R.; Guevara, J. and Castillo O. 2000. Determination of beta-globin gene cluster haplotypes and prevalence of alpha-thalassemia in sickle cell anemia patients in Venezuela. American Journal of Hematology, 64: 87-90.

Arends, A.; Chacín, M.; Bravo, M.; Montilla, S.; Guevara, J.; Velázquez, D.; García, G.; Álvarez, M. and Castillo, O. 2007. Haemoglobinopathies in Venezuela. Interciencia, 32: 516- 521.

Bavilaquia, L.; Matteri, V.; Ewald, G.; Salzano, F.; Coembra, C. and Santos, R. 1995. β- globin gene cluter haplotype distribution in the five Brazilian Indian tribes. American Journal of Physical Anthropology, 98: 395-401.

Bernal, M.; Collazos, A.; Bonilla, R. and Tascón, E. 2010. Determination of the prevalence of hemoglobin S, C, D, and G in neonates from Buenaventura, Colombia Medical, 41: 141-147.

Betancourt, J. 1995. Haemoglobin variants and haematological characteristics in patients from the population of Araya, Sucre State. Degree thesis. Department of Bioanalysis, Universidad de Oriente, Cumaná.

Brandan, N.; Aguirre, M. and Giménez, C. 2008. "Haemoglobin". "Cátedra de bioquímica".<http://www.med.unne.edu.ar/catedras/bioquimica/pdf/hemoglobin a.pdf>. (11/11/2013).

Bravo-Urquiola, M.; Arends, A.; Montilla, S.; Velásquez, D.; García, G.; Álvarez, M.; Guevara, J. and Castillo, O. 2004. Advantages of High Pressure Liquid Chromatography (HPLC-CE) in the study of haemoglobinopathies in Venezuela. Clinical Research, 45:309-315.

Bravo-Urquiola, M.; Arends, A.; Montilla, S.; Gérard, N.; Velásquez, D.; García, G.; Álvarez, M.; Guevara, J.; Castillo, O. and Krishnamoorthy, R. 2007. Molecular Characterization and origin of β thalassemia in Venezuelan Patients. American Journal of Hematology, 32: 241-248.
Calvo-Villas, J.; Zapata, M.; Cuesta, J.; De la Iglesia, S.; Ropero, P.; Carreter, E. and Sicilia,F. 2006. Prevalence of haemoglobinopathies in pregnant women in the Lanzarote health area. Annals of Internal Medicine, 25: 206-212.
Colah, R.; Surve, R.; Sawant, P.; D'Souza, E.; Italia, K.; Phanasgaonkar, S.; Nadkarni, A. and Gorakshakar, A. 2007. HPLC studies in haemoglobinopathies. Indian Journal of Pediatrics, 74: 657-662.
Cunil-Grau, P. 1987. Geografía del poblamiento venezolano en el siglo XIX. Second edition. Comisión Presidencial V Centenario de Venezuela and Universidad Central de Venezuela. Caracas Venezuela.
Cook, J.; and Monsen, E. 1976. Food iron absorption in human subjects m. Comparison of effect of animal proetins on non-heme iron absorption. American Journal clinical nuttrition. 29: 859-867.
De Galiza, G. and Da Silva, M. 2003. Molecular aspects of sickle cell anaemia. Jornal Brasileiro de Patología e Medicina Laboratorial, 6: 39-59.
De las Heras, S. and Pérez, L. 2008. Haemoglobinopathies diagnosed in the health area of the Hospital Universitario Nuestra Señora de Candelaria de Santa Cruz de Tenerife during one year. Annals of Internal Medicine, 25: 61-66.
Duran, C.; Morales, O.; Echeverrim S. and Isaza, M. 2012. Beta globin gene haplotypes in haemoglobin S carriers in Colombia. Biomédica, 32:103-11.
Evans, G.; Poulsen, R.; Canahuate, J and Pastor, J. 2003. Pregnancy-associated thalassaemia.Revista Chilena Obstetricia Ginecologica, 68:124-128.
Fairhurst, R.; Baruch, D.; Brittain, N.; Ostera, G. and Wallach, J. 2005. Abnormal display of PfEMP- 1 on erythrocytes carrying haemoglobin C may protect against malaria. Nature, 435: 1117- 1121.
García, O.; Chacín, M.; Bravo, M.; Gómez, G.; Montilla, S.; Merzón, R.; De Donato, M.; Castillo, O. and Arends, A. 2009. Diagnosis of haemoglobinopathies from heel blood of newborns in different hospital centres in Venezuela. Anales de Pediatría, 71: 314-318.
García, F.; Rodríguez, L.; Gómez, A.; Martínez, O.; Martínez, S.; González, O.; Jaime, J.; Guerra, C. and Gómez, D. 2010. Hereditary haemolytic anaemias from the perspective of a reference laboratory in Northern Mexico. Revista de Hematología, 11: 136-140.
Giardina, B.; Mosca, D. and De Rosa, M. 2004. The Bohr Effect of haemoglobin in vertebrates: an example of molecular adaptation to different physiological

requirements. Acta Physiologica Scandinavica, 182: 229-244.
nzález, M., Salazar, R., Alvarez, M., and Arends, A. 1998. Haplotypes of thegene in patients from the population of Campoma, Sucre State. Acta Cientifica Venezolana 49: 252-258.

Hempe, J. and Craver, R. 1994. Quantification of haemoglobin variants by Capillary Isoelectric Focusing. Clinical Chemistry, 40: 2288-2295.

Hoffman, R.; Benz, E.; Shattil, S.; Furie, B. and Cohen, H. 2000. Haematology: Basic Principles and Practice. Third Edition. Churchill Livingstone.

Kai, S. 2007. Instruction manual "Biophotometer plus eppendorf. Hamburg. Germany.

Kazazian, H. 1990. The thalassemia syndromes: molecular basis and prenatal diagnosis in 1990. Seminairs in Hematology, 27: 209-228.

Kunkel, H. and Wallenius, G. 1955. New haemoglobin in normal adult. Blood Science, 10:122-128.

Labie, D.; Richin, C.; Pagnier, J.; Gentilini, M. and Nagel, R. 1984. Hemoglobins S and C in Upper Volta. Human Genetics, 65: 300-302.

Lewis, S.; Bain, B. and Bates, I. 2008. Practical haematology. Tenth edition. El Sevier. Barcelona, Spain.

Long, J.; Chakravarti, A.; Bohem, C.; Antonarakis, S. and Kazazian, H. 1990. Phylogeny of human β- globin haplotypes and its implications for recent human evolutions. American Journal of Physical Anthropology, 81: 113-150.

Malcorra, J. 2001. Haemoglobinopathies and thalassaemias. Blood, 78: 2165-2177.

Martínez, G.; Hernández, L.; Muñiz, C. and Hernández, A. 1997. Molecular biology of haemoglobinopathies and malignant haemopathies. Revista Cubana Hematology Inmunology Hemoterapia, 12: 3-6.

Mata, E.; De Hollain, P. and Bauce, G. 1993. Nutritional evolution of a group of preschool children in the State of Monagas. Nutrición. 6:11-18.

Medina, J. 2013. Study of haemoglobin variants and β globin gene haplotypes in schoolchildren from the populations of Guariquen, Benítez Municipality and Petare, Bolívar Municipality. Degree thesis. Department of Biology, Universidad de Oriente, Cumaná.

Mongini, C. and Waldner, C. 1996. Methodology for the evaluation of mononuclear cells. Médica Panamericana. Argentina.

Moreno, N.; J. A. Martinez; Z. Blanco, L.; Osorio; and P. Hackshaw. Blanco, L.; Osorio. and P. Hackshaw. 2002. Beta - globin gene cluster haplotypes in Venezuelan sickle cell patients from the State of Aragua. Genetics and Molecular Biology 25: 21 - 24.

Nagel, R. 1984. The origin of the hemoglobin S gene: Clinical, genetic and anthropological consequences. Einstein Quart Journal of Medicine, 2: 53-62.

Navarrete, M. 2005. Génesis y desarrollo de la esclavitud en Colombia: siglo XVI y XVII. Colombia, Universidad del Valle.

World Health Organization (WHO). 2005. Prevalence of sickle cell disease. Meeting EB117/34. Genoa, France.

Oropeza, T.; Flores-Angulo, C.; Villegas, C.; Martínez, J.; Pulido, N.; Baeta, F. and Moreno,N. 2014. Detection of sickle cell trait carriers in a population sample from Maracay and its metropolitan area. Community and Health, 12: 46-55.

Pathrapol, L.; Amporn, L.; Tirawat, W.; Saovaros, S.; Suthat, F. and Duncan, R. 2010. A mechanism of ineffective erythropoiesis in β-thalassemia/Hb E disease. Haematologica, 95: 716-723.

Pauling, L.; Itano, H.; Singer, S. and Wells, L. 1949. Sickle cell anaemia, a molecular disease.
Science, 15: 110-543.

Peñaloza, R.; Buentello, L.; Hernández, M.; Nieva, B.; Lisker, R. and Salamanca, F. 2008. Frequency of haemoglobin S in five Mexican populations and its importance in public health. Public Health of Mexico, 50: 325-329.

Peñuela, O. 2005. Haemoglobin: a model molecule for the researcher. Colombia Medical, 36: 215-225.

Pérez, S. 2015. Hemoglobinopathies. American Journal of Hematology, 353: 498-507. Perutz, M. 1972. Hemoglobin: the molecular lung. New Science, 14: 14-18.

Rodríguez, W.; Sáenz, G. and Chaves, M. 1998. Haemoglobin S haplotypes: epidemiological, anthropological and clinical significance. Revista Panameña de Salud Pública, 3: 1-8.

Romero, C.; Gómez, A.; Durante, Y.; Amazo, C.; Manosalva, C.; Chila, L.; Casa, M. and Briceño, I. 2015. Haemoglobin variants in a population with positive diagnostic impression for haemoglobinopathies in Colombia. Revista Médica Chilena, 143: 1260-68.

Ruíz, G. 2003. Haemoglobinopathies and thalassaemia. In: Fundamentos de hematología. Ruíz- Arguelles, G. Third edition. Médica Panamericana. Argentina.

Saenz, G. 1988. Haemoglobinopathies in the countries of the Caribbean Basin. Journal of Tropical Biology, 36: 361- 372.

Sáenz, G. 2005. Abnormal haemoglobins. Acta Médica Costarricense, 47: 173-179.

Salazar, R. 1995. Haemoglobin polymorphism in the Venezuelan population. Saber, 7: 15-19.
Salazar, R.; Betancourt, J.; Brines, M. and Arends, A. 1996. Haematological characteristics and haemoglobin variants in the populations of Araya and Manicuare, Sucre State, Venezuela. Saber, 8: 50-55.
Salazar, R.; Bejarano, Y.; Gonzálcz, M. and Arends, A. 2002. Socioeconomic stratification, haematological parameters and haemoglobin variants in schoolchildren from three towns in Sucre State, Venezuela. Saber, 14: 55-59.
Salazar, R. 2004. Haemoglobin S in the Venezuelan population. Clinical Research, 45: 11-21.
Schneider, R. 1974. Differentiation of Electrophoreticall and Similar Hemoglobins-Such as 5, D, G, and P; or A2, C, E, and 0-by Electrophoresis of the Globin Chains. Clinical Chemistry, 20: 1111-1115.

Sebastiani, P.; Ramoni, M.; Notan, V.; Baldwin, H.; Steinberg, N. 2005. Genetic dissection and prognostic modeling of overt stroke in sickle cell anemia. Natural Genetics, 37:435-440.
Shimizu, K.; Hashimoto, T.; Harihara, S.; Tajima, K.; Sonoda, S. and Zaninovic, V. 2001. Beta-globin gene haplotype characteristicsof Colombian Amerindians in South America. Human Heredity, 51:54-63.
Sokal, R. and Rohlf, J. 1969. Biometry. D. W. Freeman and Co. San Francisco.
Sutton, M.; Bouhassira, E. and Nagel, R.1989. Polymerase chain reaction amplification applied to the determination of b-like globin gene cluster haplotypes. American Journal of Hematology, 32: 66-98.
Talmaci, R.; Traeger-Synodinos, J.; Kanavakis, E.; Coriu, D.; Colita, D. and Gavrila, L. 2004. Scanning of β-globin gene for identification of β-thalassemia mutation in Romanian population. Journal of Cellular and Molecular Medicine, 8: 232-240.
Tan, G.; Dunstan, S. and Lee, S. 1993. Evaluation of high perfomance liquid chromatography for Soutine estimation of haemoglobins A_2 and F. Journal of Clinical Pathology, 46: 852- 856.
Turgeon, M. 2006. Clinical haematology. Theory and procedures. Second edition. Manual moderno. Bogotá.
Vargas, C. 2011. Beta thalassaemia. Revista médica de Costa Rica y Centroamérica, 68: 355- 357.
Vera, L. 2010. Haemoglobin: a prodigious molecule. Revista Real Academia de Ciencias Exactas, Físicas y Naturales, 104: 213-232.
Villalba, T. 2015. From clinical suspicion to chance finding: evolution of

haemoglobinopathies studies. Revista Médica Chilena, 206: 1325-31.

Villalobos, A.; Bustos, R.; Casas, M.; Gutierrez, E.; Perea, F.; Thein, S.1997. β-thalassemia and β-globin gene haplotypes in Mexican mestizos. Human Genetics, 99: 498-500.

Vívenes, M.; Rodríguez-Larralde, A. and Castro- Guerra, D. 2003. Beta-globin gene cluster haplotypes as evidence of African gene flow to the northeastern coast of Venezuela. American Journal of Human Biology, 15: 1-9.

Vives, J. 2001. Anaemias due to congenital haemoglobin defects: structural haemoglobinopathies and thalassaemias. Medicine, 8: 2684-2693.

Welsh, K. and Bunce, M. 1999. Molecular typing for the MHC with PCR-SSP. Immunogen Journal, 1: 157-176.

Westermeier, R. 1997. Electrophoresis in Practice: a Guide to Methods and Applications of DNA and Protein Separation, VCH. Second edition. Weinheim. Germany.

Zafeiriou, D.2006."Neurological complication in beta-thalassemia",Brain and Development. Elsevier Science Incorporated, 20: 477-481.

METADATA SHEET

Metadata Sheet for Theses and Promotional Works - 1/6

Title	STUDY OF HAPLOTYPES OF THE HBB GENE, IN SCHOOLCHILDREN IN THE CRUZ SALMERÓN ACOSTA MUNICIPALITY, STATE OF SUCRE IN THE PERIOD 2014-2015
Subtitle	

Author(s)

Surname and first name	CVLAC code / e-mail	
EMIRATES FROM LOURDES MAZA DE FIGUEROA	CVLAC	18.211.194
	e-mail	Vivicar232010@hotmail.com
	e-mail	
	CVLAC	
	e-mail	
	e-mail	
	CVLAC	
	e-mail	
	e-mail	

Key words or phrases:

Metadata Sheet for Theses and Promotional Works - 2/6

Lines and sub-lines of research:

Area	Sub-area
Science	Biology

ABSTRACT

*A study of the frequency of haemoglobinopathies and haplotypes of the HBB gene was carried out in the towns of Araya and Manicuare, Cruz Salmerón Acosta municipality, Sucre state. Between the two populations, a total of 289 peripheral blood samples from school children were analysed. 132 peripheral blood samples were randomly collected in the Araya population and 157 samples in Manicuare. The haematological parameters (Hb, Hcto, CHCM, VCM, HCM) were analysed by an automatic haematological analyser. Peripheral blood samples were then analysed by electrophoresis on cellulose acetate membranes at pH 8.6 to identify the haemoglobin variants present in the schoolchildren, finding 6.2% (n=18) of haemoglobin variants in unrelated individuals and 93.8% (n=289) of individuals homozygous for Hb A. The samples that had variants detected were analysed by HPL-CE for confirmation, reporting the Hb S variant with a frequency of 3.8% (n=5) and the presence of Hb A2↑ with 1.5% (n=2) in the population of Araya, in the same way a frequency of Hb S (3.2%) (n=5) and Hb A2↑ with 3.8% (n=6) in heterozygous form were found in Manicuare. Once the presence of mutations in the HBB gene was confirmed in the two populations, the regions of the HBB gene cluster: 5'ε, $^{G}\gamma$, $^{A}\gamma$, ψβ and 3'ψβ were amplified and each was digested with the specific restriction enzymes HincII and HindIII. Through this analysis, it was possible to identify the characteristics of 7 haplotypes of the HBB*A gene in schoolchildren from the population of Araya, the most frequent being: haplotype 2 with 38.4%, followed by 1 and 10 (15.4%) and an atypical haplotype A (++-+-) with 7.7%. After this, 5 haplotypes of the HBB*A gene were identified in the Manicuare population, with haplotypes 2 being the most frequent (53.8%), followed by haplotypes 1, 11 and an atypical haplotype A (++-+-) with 15.4%. In the case of the HBB*S gene haplotypes identified in the Araya school population, the CAR haplotype was found with 75% and an atypical S haplotype with 25%, and in the Manicuare population the CAR and Ben haplotypes were identified (40%), followed by the Cam haplotype (20%). These results differ from those reported in the Rivero, Bolívar and Benítez municipalities and are in agreement with the results reported in the Sucre municipality. Thus, the Araya and Manicuare populations, Cruz Salmerón Acosta municipality, Sucre state, showed a particular distribution of haplotypes, finding a high frequency for the Bantu haplotype, which means that African individuals from Central Africa were dispersed throughout eastern Venezuela and, therefore, the CAR haplotype is the most frequent in Araya and, in Manicuare, Benin and CAR were found with the same proportion. Suggesting two different histories for the African slave trade and with them the distribution and origin of haemoglobinopathies in eastern Venezuela (Sucre state).*

Keywords: *haplotypes, HBB, sickle cells, sickle cell trait.*

Metadata Sheet for Theses and Promotional Works - 3/6 Contributors:

Surname and first name	ROLE / CVLAC Code / e-mail							
María José González	ROLE	CA	☐	AS	X	YOU R	JU	☐
	CVLAC							
	e-mail	mariluna1blue						
	e-mail							
Martha Bravo Urquiola	ROLE	CA	X	AS		YOU R	JU	X
	CVLAC							
	e-mail							
	e-mail							
Merlyn Vivenes de Lugo José Gregorio Betancourt	ROLE	CA	☐	AS		YOU R	JU	X
	CVLAC							
	e-mail							
	e-mail							

Date of discussion and approval: Year Month Day

2016	07	22

Language:**SPA**

Metadata Sheet for Theses and Promotional Works - 4/6

File(s):

File name	MIME type
Thesis- Emirjes De Lourdes Maza De Figueroa.doc	Application/Word

Scope:

Space :	National	(Optional)
Temporary:	Temporary	(Optional)

Degree or diploma associated with the job:

Degree in Biology

Level Associated with Work:

Bachelor's degree

Study Area:

Biology

Institution(s) guaranteeing the degree(s):

University of Oriente Núcleo de Sucre

Hoja de Metadatos para Tesis y Trabajos de Ascenso – 5/6

UNIVERSIDAD DE ORIENTE
CONSEJO UNIVERSITARIO
RECTORADO

CU N° 0975

Cumaná, 04 AGO 2009

Ciudadano
Prof. JESÚS MARTÍNEZ YÉPEZ
Vicerrector Académico
Universidad de Oriente
Su Despacho

Estimado Profesor Martínez:

Cumplo en notificarle que el Consejo Universitario, en Reunión Ordinaria celebrada en Centro de Convenciones de Cantaura, los días 28 y 29 de julio de 2009, conoció el punto de agenda ***"SOLICITUD DE AUTORIZACIÓN PARA PUBLICAR TODA LA PRODUCCIÓN INTELECTUAL DE LA UNIVERSIDAD DE ORIENTE EN EL REPOSITORIO INSTITUCIONAL DE LA UDO, SEGÚN VRAC N° 696/2009".***

Leído el oficio SIBI – 139/2009 de fecha 09-07-2009, suscrita por el Dr. Abul K. Bashirullah, Director de Bibliotecas, este Cuerpo Colegiado decidió, por unanimidad, autorizar la publicación de toda la producción intelectual de la Universidad de Oriente en el Repositorio en cuestión.

Comunicación que hago a usted a los fines consiguientes.

UNIVERSIDAD DE ORIENTE
SISTEMA DE BIBLIOTECA
RECIBIDO POR
FECHA 5/8/09 HORA 5:30

Cordialmente,

JUAN A. BOLAÑOS CURVELO
Secretario

REPÚBLICA BOLIVARIANA DE VENEZUELA
UNIVERSIDAD DE ORIENTE
SECRETARÍA
CONSEJO UNIVERSITARIO

C.C: *Rectora, Vicerrectora Administrativa, Decanos de los Núcleos, Coordinador General de Administración, Director de Personal, Dirección de Finanzas, Dirección de Presupuesto, Contraloría Interna, Consultoría Jurídica, Director de Bibliotecas, Dirección de Publicaciones, Dirección de Computación, Coordinación de Teleinformática, Coordinación General de Posgrado.*

JABC/YOC/maruja

Apartado Correos 094 / Telfs: 4008042 - 4008044 / 8008045 Telefax: 4008043 / Cumaná - Venezuela

Hoja de Metadatos para Tesis y Trabajos de Ascenso – 6/6

Artículo 41 del REGLAMENTO DE TRABAJO DE PREGRADO (vigente a partir del II Semestre 2009, según comunicación CU-034-2009): "Los trabajos de grados son de la exclusiva propiedad de la Universidad de Oriente, y solo podrá ser utilizados para otros fines con el consentimiento del Concejo de Núcleo respectivo, quien deberá participarlo previamente al Concejo Universitario, para su autorización".

Emirjes De Lourdes Maza De Figueroa
AUTOR

María José González
TUTOR

More Books!

yes

I want morebooks!

Buy your books fast and straightforward online - at one of world's fastest growing online book stores! Environmentally sound due to Print-on-Demand technologies.

Buy your books online at
www.morebooks.shop

Kaufen Sie Ihre Bücher schnell und unkompliziert online – auf einer der am schnellsten wachsenden Buchhandelsplattformen weltweit! Dank Print-On-Demand umwelt- und ressourcenschonend produzi ert.

Bücher schneller online kaufen
www.morebooks.shop

info@omniscriptum.com
www.omniscriptum.com

OMNIScriptum

MIX
Papier aus verantwortungsvollen Quellen
Paper from responsible sources
FSC® C105338

Printed by Books on Demand GmbH, Norderstedt / Germany